Earl C. Haag teaches English and German, including Pennsylvania German, at Penn State's Schuylkill Campus. Professor Haag developed his interest in Pennsylvania German while doing graduate study of Palatinate dialects under Buffington at Penn State and in Germany at Heidelberg University.

A Pennsylvania German Reader and Grammar

FER MEI MUDDER,

wu mir immer gsaat hot,

ich sett mich draahalde.

A Pennsylvania German Reader and Grammar

Earl C. Haag

Keystone Books
The Pennsylvania State University Press
University Park and London

Library of Congress Cataloging in Publication Data

Haag, Earl C.
A Pennsylvania German reader and grammar.

(Keystone books)
Text in English and German.
1. Pennsylvania German dialect.   2. German
language--Grammar.   3. German language--Readers.
I. Title.   II. Series.
PF5934.H3   437'.9748   82-80453
ISBN 0-271-00314-6   AACR2

TABLE OF CONTENTS

viii

FOREWORD

Dialect speakers are, in the main, an egotistical
lot. They tend to think of their versions of the dialect grammar and vocabulary as the best forms--if not,
indeed, the only correct forms. For this and several
other reasons, any attempt to set a dialect down on
the printed page is fraught with danger. No dialect
writer can keep all of his readers happy all of the
time.

For example, let us assume that we are attempting
to reproduce in print the results of running together
the two words "have them." Should we write "have em"
or "have um"? Will those speakers of the language who
tend toward "have im" be unhappy? What about "hav em"
("hav um," "hav im")? Or should we use an apostrophe
to indicate missing letters: "hav 'em" ("hav 'um,"
"hav 'im")? Shouldn't we actually run the words together on the printed page, too: "havem" ("havum,"
"havim")?

More complicated would be the rendering of "What do
you know?" Some possibilities are "What do yuh know?"
"Wha do yuh know?" "Wha duh yuh know?" and "Whaduhyuh-
know?" The reader can no doubt think of other possibilities, such as "Wadayaknow?" Which form is the correct one? Or are they all correct? If they are all
correct, which one do we choose? If we choose one,
will someone insist we should have chosen another? And
finally, should we use the letter k at all? It's not
pronounced, so why use it in our printed rendition?
These examples are, of course, in English, but they
illustrate an important problem in writing a dialect:
dialects, for the most part, have no standard, codi-
fied spelling and grammar rules used by all members of
the dialect group.

Thus I have found the following variations of the
Pennsylvania German (PG) equivalents of "these days,"
or "presently," in PG literature: heidzudaags, heidzedaags, heidesdaags, heidedaags, and heidichdaags. I
have noted all but the last written with t in place of

ix

<u>d</u>, e.g., <u>heitzedaags</u>. A caller on Dave Hendricks'
radio program, "Die Wunnerfitz Schtunn," used <u>heides-
daags</u> and <u>heitzudaags</u> in two successive sentences! Is
one of these variants "more correct" than any other?
I have listed many variants in the notes that accompany chapter vocabulary lists, but I know I run the
risk of aggravating a reader who has never heard of a
particular variant or whose favorite variant I have
not included.

In the case of PG, a second problem concerns the
sound values given the letters used to spell the dialect. Many PG writers, both past and present, have
opted for the English sound values to spell the dialect. As early as 1879 E.H. Rauch wrote in his <u>Penn-
sylvania Dutch Handbook</u>: ". . . I have very much im-
proved and simplified the spelling, which is strictly
according to English rules . . . Anyone who can read
English, can also read Pennsylvania Dutch as I have
recorded it. . . . To read it, no study of orthography is at all necessary, because it is simply English."

In effect, the reader uses his knowledge of English
sounds (especially vowel sounds) to read the PG word.
Several present-day writers of PG columns published in
newspapers circulating in PG areas use this system,
and their regular readers, I am sure, have become ac-
customed to the system and would not want the writers
to change.

But there is also the German system of sound values
to be considered. In his introduction to a collection
of PG poetry and prose, entitled <u>Pennsylvania German</u>
(1904), Daniel Miller wrote he considered it unfortu-
nate that PG orthography "has not been determined from
the standpoint of the grammatical German, so as to
secure uniformity in the modes of writing." And in
his introduction to the second volume (1911) he wrote,
"The dialect is German, not English, and all attempts
to present it in English form will do violence and in-
justice to it, and fail to secure honor for it." This
latter thought can be debated, I should think, but one
irrefutable fact has led me to use the German system
of sound values, and thus the German system of

orthography, for this book: because English vowels
have as many as four or five sound values, the reader
must have a prior knowledge of the dialect to read PG
spelled "in English." Only those readers who already
know the dialect can be certain they are pronouncing
the PG word correctly.

For example, Joseph H. Light, in his Der Alt Schul-
meeschter (1928) spells the PG word for "house" as
hous, with ou pronounced as in E "house." He spells
the PG word for "just" as youscht, with the ou this
time pronounced like ou in E "you." One more example
taken from scores of such doublets: PG kott "had,"
with o pronounced as in E "hot," but PG soll with o
pronounced (approximately) as in "go."

Thus my conclusion that a text meant for the beginning student of the dialect must use German orthography; if nothing more, the relatively controlled,
"purer" sound values of German vowels allow the student
a much better chance at an informed guess of the correct pronunciation of dialect words he has not seen
before.

But this does not mean that all problems associated
with pronunciation have been solved. We most often
teach the sounds of a foreign language by citing English equivalents. However, one's pronunciation of an
English vowel is often dependent upon the area in
which he was raised or even influenced by his knowledge
of the foreign language. For example, often a PG text
tells us that the o in PG Kopp is pronounced like the
u in E "cup." It would appear that the PG speaker's
pronunciation of "cup" has been influenced by the pronunciation of PG Kobbche or Kobbli "cup." But this is
certainly a relatively minor problem (especially if
someone is studying PG with the help of a teacher,
relative, or friend who knows the dialect), detracting
but little from the benefits gained by using German
orthography to write PG.

The reader will soon realize that this book is not
exactly like other language books, the books from
which he learned German, Spanish, or French in high
school or college. It is, as the title implies, a

reader as well as a grammar. As such, I felt it my
duty to include most of the reasonably common vocabulary for the subject matter of each chapter. Many of
my students who spoke PG even before attending classes
have told me that the text includes many words they
have never used or heard of. I am then glad I have
given them the opportunity to gather up some of the
words and expressions that have been lost during the
three hundred years the Pennsylvania Germans have been
in America.

I could add that I hope this book will prepare the
reader to go on to bigger and better things: the rich
and varied--and actually quite extensive--PG literature. My dream, you see, is that one day in the near
future I will be able to walk into book stores in the
PG areas of Pennsylvania, if not throughout the entire
state, and pick a book of PG literature off the
shelves.

But be that as it may, should an instructor choose
to use this book in a PG class, he could easily limit
the number of words to be memorized by the class. I,
myself, do this by telling my students they must know
at least the vocabulary in the translation exercises
that accompany the list of Minimum Essentials, as I
call them, which I hand out for every chapter. By the
way, the student finds the English sentences translated into PG on the back of the hand-out, just as the
reader of this book finds English translations on the
right-hand pages of the reading selections. As a student, I wasted too many hours in foreign language
classes looking up words in the back of the book, and
long ago came to the conclusion that I wasn't learning
anything while looking them up; only after I had found
them could I begin to learn.

And speaking of English translations, I hope readers who are well versed in formal English will forgive
my less-than-formal and unidiomatic English constructions they will at times find in the translations of
the PG texts. Especially for the beginning student, I
have attempted to translate the PG as literally as
possible without duplicating the ridiculous restaurant
place-mat translations like "Throw the cow over the

fence some hay." Thus, "The schoolmaster is also sitting already at the desk" may not be idiomatic English,
but it does reflect for the beginning student the pro-
gression of words in the comparable PG construction, a
result that I believe justifies sacrificing a perfectly idiomatic English equivalent.

Another difference lies in the fact that the reading and grammar sections do not always cover the same
number of pages in each chapter and thus do not always
take the same amount of time to complete. In fact,
some of the chapters toward the end of the book have
no grammar sections at all. My reason for this
"irregularity" is that as students become more and
more proficient in a language, they spend more and
more time practicing the spoken language and looking
at visual aids such as slides and motion pictures. It
is reasonable to shorten text material to allow time
for other endeavors.

Furthermore, the scholar will find neither noun
classes nor verb classes in this book. During my
twenty-five years of teaching Standard German classes
I have never known a student to memorize nouns or
verbs by class. Students simply learn the plurals of
nouns and the principal parts of verbs as they come up
in chapter vocabularies. When it comes to using a
language, knowledge of such classes is at best merely
interesting. How many Germans or Pennsylvania Germans,
though they be highly educated and/or though they
speak their language beautifully, think of their nouns
and verbs in terms of classes? Were this book meant
to be just another reference grammar, I would, of
course, have included such classes.

My format of two parts for each chapter (with three
exceptions) is also based on my years of experience in
the classroom. The complaints I have heard most often
in my college foreign language classes are that the
text moves too quickly, that the students don't get a
chance to "catch their breath," that there is no
opportunity to review and immediately reinforce the
learning that took place in a previous chapter. But
when using this text, the students who take the time
to master Part One of a given chapter will find the

grammar and/or vocabulary of Part Two relatively easy, resulting in a psychological boost not enjoyed by students studying foreign languages in today's college and high school textbooks.

Let me at this point add something about my use of English loan words. Many scholars of the PG dialect would like to keep the dialect pure; that is, they would like to eliminate English loan words and replace them either with suitable, but long neglected, PG words, or even with St G words duly "doctored" to conform to PG patterns. The only E loan words I have used in this book are those I found regularly in the PG literature of the early years (before 1900), the middle years (1901-1940), and the modern years (1941-present day) as well as have personally heard used by a native PG speaker. The preceding classification is, admittedly, arbitrary, but I believe it serves my purposes well: finding a loan word in the literature of all three periods as well as hearing the word personally proves a consistency that cannot be ignored. Such loan words (_Fens_ "fence" and _Tietscher_ "teacher" are two obvious examples) have become so firmly entrenched in the dialect that they cannot be deleted out of hand from the PG vocabulary.

Lack of space dictated that the exercises I ask my students to do be limited to the few found as examples in the appendix. I call them practice patterns, and I believe in them. They differ from the exercises usually found in language texts in that the latter almost always do nothing but test whether the student has already learned the material, but the former are designed to help the reader learn the material. My experiences with these practice patterns has been excellent; they work!

At this point, I wish to acknowledge the role played by Bill Zimmerman, Assistant Director for Continuing Education at the Schuylkill Campus of The Pennsylvania State University, who first suggested that I develop a Pennsylvania German language and culture course, the result of which is this book. Bill believed in me, and I needed that as I began this new

endeavor. I also thank the College of the Liberal
Arts, The Pennsylvania State University, whose grants
made possible the final typing of the manuscript.

The epigraphs gleaned from Pennsylvania German Society and Pennsylvania German Folklore Society publi-
cations have been used with the kind permission of
Pastor Frederick S. Weiser, editor for the Pennsylvania German Society.

Earl C. Haag

The Pennsylvania State University
Schuylkill Campus
Schuylkill Haven, Pa. 17972

January 1982

SOURCES OF EPIGRAPHS

'S Alt Marik-Haus Mittes in D'r Schtadt, un Die Alte
Zeite. York: H.L. Fischer, 1878.

Am Schwarze Baer. Philadelphia: Pennsylvania German
Folklore Society, 1947.

Bilder un Gedanke. Breinigsville: Pennsylvania German Society, 1975.

Drauss un Deheem. Philadelphia: Pennsylvania German
Folklore Society, 1936.

Gezwitscher. Philadelphia: Pennsylvania German Folklore Society, 1938.

Harbaugh's Harfe. Philadelphia: Reformed Church Publication Board, 1870.

The Later Poems of John Birmelin. Philadelphia: Pennsylvania German Folklore Society, 1951.

Pennsylvania German, 2nd ed. Reading: Daniel Miller,
1904.

Pennsylvania German, Vol. II. Reading: Daniel
Miller, 1911.

The Pennsylvania-German Dialect. Lancaster: Pennsylvania German Society, 1902.

Pennsylvania German Manual. Kutztown: Urick and
Gehring, 1875.

Pennsylvania German Verse. Norristown: Pennsylvania
German Society, 1940.

PENNSYLVANIA GERMAN SOUNDS

Letter(s)	E equivalents	PG Examples
Short Vowels		
a	o in lot, got	alt, old; nass, wet; Karrich, church
ae	a in mat, fat	Maetsch, match; Kaerrich, church
e	e in met, wet	nemme, take; zwett, second
i	i in sit, fit	Kind, child; finne, find
o	o in hoe, but without the diphthongal glide to u	Rock, coat; oft, often
u	oo in foot	Sunn, sun; Mudder, mother
Long Vowels		
aa or aah	aw in law, saw	Fraa, wife; Baam, tree
e,ee,eeh eh	ai in fail	lese, read; fehle, to be missing
ae	a in has	schwaer, heavy; gaern, gladly
ih,ie,ieh	ie in believe	dihr, to you; Fiess, feet; frieh, early
o,oo,oh	o in go without the diphthongal glide to u	rot, red; grooss, big; froh, happy
u,uu,uh	oo in food	zu, to,closed; Fuuss, foot; Schtuhl, chair

Letter(s)	E Equivalents	PG Examples
Diphthongs		
au	ou in house, mouse	aus, out; brauche, need
ei	ie in pie	drei, three; mei, my
oi	oi in oil	Oi, egg; Moi, May
Two Vowel-Like Sounds		
final -er	a in father	Vadder, father; Mudder, mother
r between vowel and consonant, or diphthong and consonant, or in final position	a in father	fiehrt, lead; mir, to me; Ohr, ear
Consonants		
b	like E b, but like p when final	binne, to tie; Kalb, calf
ch	after front vowels like whispered k; after back vowels like clearing your throat	Dach, roof; Decher, roofs; Loch, hole; Lecher, holes
d	like E d, but like t when final	Daag, day; Feld, field
f	like E f	froh, happy; uff, up, on

Letter(s)	E Equivalents	PG Examples
g	g in get, but like k when final; when medial, much like y in canyon after front vowels; almost inaudible after back vowels	leige, to lie (recline); Daage, days
h	like E h, unless medial or final, when it is not pronounced, but indicates preceding vowel is long	halb, half; hawwe, have; gehne, go; Kuh, cow
k,ck,(gh)	k in kite	kalt, col; wacker, awake; (gh of past participles: gheere, heard; ghockt, sat)
l	like E l	
m	like E m	
n	like E n	
ng	like ng in singer, never like ng in finger	singe, to sing; Finger, finger
p,(bh)	like E p	Peif, pipe; bhalde, to keep (be- + h = bh)
r	E r trilled or flapped once	rufe, call; Biere, pears
s,ss	like E s	
sch	sh in shut	schee, nice; wesche, wash
tsch	ch in church	Tschanni, Johnny; deitsch, German

Letter(s)	E Equivalents	PG Examples
t	E t (but changes to d(d) between vowels	Tee, tea; Schtarrick, strong; Zeit, time (but Zeide, times); alt, old (but elder, older)
v	like E f	vier, four; vazeh, fourteen
w or ww	like E v	was? what?; gewwe, give
x	x in six	nix, nothing; Ox, ox
y	y in yes	ya, yes; Yaahr, year
z(tz)	ts in nets, rats	zwee, two; Zeit, time; Katz, cat

NOTE: Pronunciations may vary from one PG area to another. If possible, consult a "native speaker" who lives in your area.

PART ONE

CHAPTER ONE

Part One

We begin our study of Pennsylvania German (PG) with
School, not because school is of first importance, but
because most students find themselves in a schoolroom
situation, and can thus make good use of it to start
their studies. Also, many of the words in the vocabulary of this first lesson--such as wall, floor, window,
door--can be "taken home" and used in everyday situa-
tions.

> *Die Maeri hot en Lemmel ghatt,*
> *Sei Woll waar weiss wie Schnee,*
> *Un wu die Maeri hiegange iss,*
> *Des Lamm waar schuur zu geh.*
> *Es iss emol mit noch der Schul*
> *Sei Gschpichde datt zu mache,*
> *Noh hen die Kinner in der Schul*
> *Aagfange laut zu lache.*[1]

[1] *David B. Brunner, from "Die Mary Hot En Lemmel
Ghatt," Pennsylvania German Verse.*

Bibliographic information
for epigraphs is shown on
p. xvi.

3

4

Der Aerscht Schuldaag (Net So Lang Zrick)

Die Schulbell ringt. Die Schulkinner gehne ins (in+es) Schulhaus. Es Schulhaus hot yuscht ee Schtubb. Die groosse Kinner hocke links. Die gleene Kinner hocke rechts. Sie hocke uff Benk un Schtiehl. Der Schulmeeschder sitzt am (an+em) Schreibdisch.

Er schteht uff un saagt, "Gude Marriye, Klass. Heit iss der aerscht Schuldaag. Ich hab viel Naame do. Ich kenn deel schunn(t). Wu iss die Palli Dreisbach?"

Die Palli schteht uff un saagt, "Ich bin do."

Der Meeschder froogt, "Schwetscht du Englisch?"

"Ya, Tietscher, gewiss. Ich kann Englisch schwetze, lese, un schreiwe."

"Un kann dei gleener Bruder, der Tschanni, Englisch schwetze?"

"Nee, er kann net. Awwer er lannt schnell."

Der Tietscher froogt annere Kwestschyens: "Waer bischt du?" odder "Wie heescht du?" odder "Was iss dei Naame?" Er froogt aa, "Wu wuhnscht du?" odder "Wu kummscht du haer?"

Deel Kinner kenne Englisch schwetze, annere kenne yuscht Deitsch schwetze. Deel wuhne uff em Land, annere wuhne im (in+em) Schteddel. Deel sin gscheit un wisse un verschtehne alles; annere sin dumm un wisse nix.

Die Schulschtubb hot vier Wend. Der Bodde(m) iss unne un die Deck iss owwe. Vanne iss es Schwazbord (odder Blaeckbord) un en Dier. Die Uhr hengt aa datt (dadde). Die Wand rechts hot drei Fenschdere. Die Wand links hot zwee Fenschdere un ee Dier.

Der Meeschder schreibt uffs Blaeckbord mit Greid. Er yuust Fedder un Dinde uff Babier. Die groosse Kinner kenne en Bleibensil yuuse. Sie lese alle Daag. Sie hen en Schpellingbuch un aa annere Bicher.

The First School Day (Not So Long Ago)

The schoolbell rings. The school children go into the schoolhouse. The schoolhouse has just one room. The big children sit (to the) left. The little children sit to the right. They sit on benches and chairs. The schoolmaster sits at the desk.

He stands up and says, "Good morning, class. Today is the first school day. I have many names here. I know some already. Where is Polly Dreisbach?"

Polly stands up and says, "I am here."

The (school)master asks, "Do you speak English?"

"Yes, teacher, of course. I can speak, read, and write English."

"And can your little brother, Johnny, speak English?"

"No, he can not. But he learns fast."

The teacher asks other questions: "Who are you?" or "What are you called?" or "What is your name?" He also asks, "Where do you live?" or "Where do you come from?"

Some children can speak English, others can just speak German. Some live in the country, others live in town. Some are smart and know and understand everything; others are stupid and know nothing.

The schoolroom has four walls. The floor is below and the ceiling is above. In front is the blackboard (loan word) and a door. The clock hangs there too. The wall to the right has three windows. The wall left has two windows and one door.

The master writes on the blackboard with chalk. He uses pen and ink on paper. The big children can use a lead pencil. They read every day. They have a spelling book and also other books.

Vocabulary

      aa: also, too
      aerscht: first
      all(e): all, every   alle Daag: every day
      alles: everything
      am (an+em): at, to the (dative)
      an: at, to
      annere: other(s)
      awwer: but
  es Babier', Babier'e: paper
 die Bank, Benk: bench
 die Bell, Bells (Belle): bell
 der Bensil, Bensils: pencil
  es Blaeckbord, -borde: blackboard
 der Bleibensil, -bensils: lead pencil
 der Bodde(m): floor
 der Bruder, Brieder: brother
  es Buch, Bicher: book
 der Daag, Daage: day
      dann: then
      datt: there
 die Deck, Decke: ceil____.
      deel: some
      Deitsch: German
 die Dier, Diere: door
 die Dinde: ink
 der Disch, Dische: table
      do: here
      drei: three
      dumm: stupid
      ee: one
 die Fedder, Feddere: pen, feather
  es Fenschder, Fenschdere: window
      frooge: to ask
      geh: to go
      gewiss: of course, certainly
      glee (gleen+ending): small
 die Greid: chalk (crayon)
      grooss: large, big
      gscheit: smart
      gut: good
      heesse: to call, be named, be called
      heit: today
      henge: to hang

```
    hocke:  to sit
    in:  in, into
    ins (in+es):  into the
    kenne:  to be able to, can
    kenne:  to be acquainted with
 es Kind, Kinner:  child
die Klass, Klasse:  class
    kumme:  to come
 es Land, Lenner:  land, country;  uff em Land:  in
        the country
    lanne:  to learn
    lese:  to read
    links:  (to the) left
    marriye:  tomorrow, morning
der Meeschder, Meeschder:  master
    mit:  with
der Naame, Naame:  name
    nee:  no
    net:  not
    nix:  nothing
    odder:  or
    owwe:  above
    rechts:  (to the) right
    ringe:  to ring
    saage:  to say
    schnell:  fast, quickly
 es Schpellingbuch, -bicher:  spelling book
der Schreibdisch, -dische:  desk
    schreiwe:  to write
 es Schteddel, Schteddelcher (Schteddel):  town,
        village
    schteh:  to stand
die Schtubb, Schtuwwe or Schtubbe:  room
der Schtuhl, Schtiehl:  chair
die Schul, Schule:  school
    schunn(t):  already
 es Schwazbord, -borde:  blackboard
    schwetze:  to talk
    sei:  to be
    sitze:  to sit
der Tietscher, Tietscher:  teacher (m.)
    uff:  on, upon
die Uhr, Uhre:  clock, watch
    un:  and
```

```
      unne:  below
      vanne:  in front
      verschteh':  to understand
      viel(e):  many, much
      vier:  four
      waer?:  who?
die Wand, Wend:  wall
      was?:  what?
  es Watt, Wadde:  word
      wie?:  how?
      wisse:  to know (a fact)
      wu?:  where?
      wuhne:  to live, reside
      ya:  yes
      yuscht:  just
      yuuse:  to use
      zwee:  two
```

Vocabulary Notes

You have no doubt noticed that many words in the text
and in the vocabulary list are capitalized even
though they are neither the first words of sentences nor proper nouns. In keeping with the Buffing-
ton/Barba system, all words used as nouns are
capitalized.

As you study the vocabulary, be sure to memorize the
definite article (E "the," PG der, die, es) that
precedes each noun. It is the article that tells
you whether a noun is masculine der, feminine die,
or neuter es. This is especially important because
PG nouns do not have obvious gender: der Leffel
"spoon," die Gawwel "fork," es Messer "knife."
Also as we shall see in future chapters, endings on
various noun modifiers will depend on the gender of
the nouns.

Almost all PG nouns, like almost all E nouns, have
both singular (one) and plural (two or more) forms.
The vocabulary lists of the first three chapters
will give the entire plural form as well as the
singular form: der Mann, Menner. All PG plural
nouns have the definite article die in the

nominative case, and it is therefore usually deleted from word lists.

Generally, the first syllable of a PG word is accented: Bruder "brother," Daage "days," hocke "to sit," annere "others." This is normally true even if the noun is a compound made up of two or more nouns, e.g., Schulbell "school bell," Schreibdisch "desk," or even a modifier and a noun, e.g., Blaeckbord "blackboard."

The only noun exception in the preceding reading selection is Babier (Fabier) "paper," originally a foreign (non-German) word. A verb exception to this rule is verschteh "to understand," which is accented on the stem of the verb, schteh. In following vocabularies, an accent mark (') will be placed after accented syllables whenever the accent does not follow the general rules for PG pronunciation.

die Deck is often replaced with the E loan word die Sieling.

der Desk (Dest) are very popular loan words for Schreibdisch, which literally means "writing table."

saage has an irregular 3rd person, singular, form secht, used by a minority of PG speakers, but it is found quite often in PG literature.

schnell: if motion is implied, then the most common PG word for "fast" is schtarrick.

verschteh "to understand," is conjugated exactly like schteh "to stand." Ver- is one of four common prefixes (be-, er-, ge-, ver-) that change the meaning of a verb, but do not change the conjugation of the verb. These four prefixes are never separated from the verb, and are never accented; the accent remains on the stem of the verb.

-o0o-

NOTE: Once again, I ask the reader to forgive the occasional lack of idiomatic English translations, the result of my desire to render the PG into English equivalents that reflect as closely as reasonable the PG constructions.

Grammar

A. The Definite Article

1. English has but one definite article, "the," no
matter what the number (singular or plural) or the
gender (masculine, feminine, or neuter). PG, however,
has the following forms of the definite article for
the nominative and accusative cases:

	Singular		
	Masculine (M.)	Feminine (F.)	Neuter (N.)
Nominative (N.) and Accusative (A.)	der	die	es
	Plural--All Genders		
N. and A.	die		

2. The nominative case is the case of the subject
of a verb:

Der Meeschder froogt...	The teacher asks...
Die Schulbell ringt...	The schoolbell rings...
Es Schulhaus hot...	The schoolhouse has...
Die Schulkinner gehne...	The school children go...

3. The nominative case is also the case of a noun
that follows a verb like sei "to be" (a linking verb):

| Er iss der Bruder. | He is the brother. |
| Sell iss es Buch. | That is the book. |

Other linking verbs will be pointed out as they appear
in future chapters.

4. The accusative case is the case of the direct
object of a verb:

| Der Bu hot es Buch. | The boy has the book. |
| Der Tietscher froogt
die Kinner. | The teacher asks the
children. |

Es Buch is the direct object of the verb hot and die
Kinner is the direct object of the verb froogt.

5. The accusative case is also the case of various other objective forms that will be cited as they come up in subsequent chapters.

6. The dative case will be taken up in the next reading selection.

7. By the way, perhaps you noticed that PG speakers like to use the definite article before proper names:

Die Palli iss schunnt   [The] Polly is already
do, awwer der Tschanni here, but [the] Johnny
iss net.   is not.

B. The Indefinite Article

1. English has two indefinite articles: "a," used before a word beginning with a consonant, and "an," used before a word beginning with a vowel (a,e,i,o,u). The PG indefinite article is en for all genders in the nominative and accusative cases. In the sentence

En Leffel, en Gawwel,   A spoon, a fork, and a
un en Messer sin do.   knife are here.

Leffel, Gawwel, and Messer are subjects (nominative case) of the verb sin. And in the sentence

Ich hab en Leffel, en   I have a spoon, a fork,
Gawwel, un en Messer.   and a knife.

these same three nouns are direct objects (accusative case) of the verb hab.

There are, of course, no plural forms of "a," "an," or en.

2. The dative case of the indefinite article will be taken up in the next reading selection.

C. The Personal Pronouns

1. The E personal pronouns are "I," "you," "he," "she," "it" in the singular; and "we," "you," "they" in the plural. The PG equivalents are the following:

	Singular		
	Accented	Unaccented	
1st Person	ich	ich	(I)
2nd Person	du	d(e)	(you)
3rd Person			
Masculine	aer	er	(he)
Feminine	sie	se	(she)
Neuter	es	(e)s	(it)
	Plural		
1.	mir	mer	(we)
2.	dihr,ihr,	der,er,	
	nihr	ner	(you)
3.	sie	se	(they)

2. The unaccented forms are used very frequently in PG. An E equivalent would be the pronoun "you," which often appears as an unaccented "yuh" in questions: "Are yuh goin'?" Indeed, the E speaker may run verb and pronoun together: "Aryuh goin'?" This happens very frequently in PG, especially if the pronoun follows the verb. In the next reading selection, for instance, Polly says, Do binnich (bin ich) and the teacher says, Do hawwich (hab ich) die Naame.

D. Verbs, Present Tense

1. With few exceptions, the formation of the present indicative of PG verbs proceeds by dropping the ending -e from the infinitive and adding endings to the resulting stem. For example, if we drop the infinitive ending -e from kumme, we then have left the stem of the infinitive, kumm, to which we add the following personal endings:

Singular	Plural
1. - (no ending)	1. -e
2. -scht	2. -e, -t, or -et
3. -t	3. -e

These verb forms thus result:

Singular	Plural
1. ich kumm (I come)	1. mir, mer kumme (we come)
2. du kummscht (you come)	2. dihr,der kummt; dihr, der kumme; ihr,er kumme; ihr,er kummt; nihr,ner kumme (you come)
3. er,sie,es kummt (he, she,it comes)	3. sie kumme (they come)

2. The second person plural has several forms, each of which is common to a particular area where PG is spoken. If you are already familiar with one of the forms, use it. If you know someone who speaks PG, ask him which form he uses and use it yourself.

3. The verbs that appear in the vocabulary of the preceding reading selection and that follow the model kumme are henge, hocke, kenne, lanne, ringe, and wuhne.

4. If an infinitive stem ends in a sibilant (s, ss, sch, or z), then the sibilant is dropped before adding the second person singular ending -scht. For example, instead of du schwetzscht, we write du schwetscht. In the preceding reading selection, heesse, lese, sitze, and yuuse also follow this pattern.

Other such verbs will be pointed out as they appear in future vocabulary lists.

5. Whenever a PG infinitive stem ends in w, this letter is retained only if it is followed by the personal ending -e. Otherwise, the w changes to a b. An example is schreiwe "to write."

Singular	Plural
1. ich schreib	1. mir,mer schreiwe
2. du schreibscht	2. dihr,der schreibt; ihr,er schreiwe, etc.
3. er,sie,es schreibt	3. sie schreiwe

Other verbs whose conjugations are like that of schreiwe will be pointed out as they appear in the vocabulary lists.

6. Two PG verbs used in the preceding reading selection, geh "to go" and schteh "to stand," are irregular in that an -n is added to the stem before the personal ending -e.

The present tense forms of geh, which may serve also as a model for schteh, are as follows:

Singular	Plural
1. ich geh	1. mir,mer gehne, gehn
2. du gehscht	2. dihr,der geht; ihr,er gehne or gehn
3. er,sie,es geht	3. sie gehne, gehn

PG speakers in Lehigh and surrounding counties use gehn (schtehn) as the plural form of this verb.

7. Two PG verbs have seemingly nothing to do with other verbs and should be memorized separately.

a) sei "to be"

Singular	Plural
1. ich bin	1. mir,mer sin
2. du bischt	2. dihr,der seid; dihr,der sin; dihr,der sind; ihr, er seid; ihr,er sin; nihr,ner sin; dihr,der seid; dihr,der sind
3. er,sie,es iss	3. sie sin

b) wisse "to know (a fact)"

Singular	Plural
1. ich weess	1. mir,mer wisse
2. du weescht	2. dihr,der wisst; ihr,er wisse; etc.
3. er,sie,es weess	3. sie wisse

8. Kenne "to be able to," "can," is an example of a group of verbs called modal auxiliaries (e.g., E can, may, or must), all of which will be presented in detail in Chapters 5 and 6. But kenne is introduced in this lesson because one naturally uses it to ask someone whether he can do something, or to state that one can do something.

Singular	Plural
1. ich kann	1. mir,mer kenne
2. du kannscht	2. dihr,der kennt; ihr,er kenne, etc.
3. er,sie,es kann	3. sie kenne

9. Any verb used with <u>kenne</u> or any other modal auxiliary appears in its infinitive form: <u>ich kann</u> <u>schwetze</u>, <u>du kannscht schwetze</u>, etc. Details concerning word order for a modal auxiliary and a dependent infinitive will be given in Chapter 5, but briefly stated, various objects and modifiers must be "sandwiched" between the modal and the infinitive:

Ich kann yuscht Deitsch   I can speak only German.
schwetze.

10. The modal <u>kenne</u> "to be able to" must not be confused with the regular verb <u>kenne</u> "to be acquainted with someone or something," whose conjugation in the present tense follows:

Singular	Plural
1. ich kenn	1. mir,mer kenne
2. du kennscht	2. dihr,der kennt; ihr,er kenne; etc.
3. er,sie,es kennt	3. sie kenne

Ich kenn der Meeschder.   I know (am acquainted with) the schoolmaster.

11. But remember, <u>wisse</u> is used when a <u>fact</u> is known:

Ich weess sell.   I know that.
Ich weess, der Meesch-   I know (the fact that) the
der iss schunn do.   teacher is already here.

E. Word Order

1. The entire first reading selection of this lesson is made up of sentences containing normal word order; that is, the subject (and its modifiers, if any) stands in first position, the verb comes next, in second position, and the rest of the sentence follows.

Die Schulbell   ringt.
subject   verb

Es Schulhaus   hot   ee Schtubb.
subject   verb   direct object

Die Schulkinner   gehne   ins Schulhaus.
subject   verb   prepositional phrase

Die gleene Kinner   hocke   rechts.
subject   verb   adverb

Just as in English, then, the PG verb is normally the second element in a declarative sentence (clause).

2. However, the subject and verb are "inverted" for questions:

Gehscht du in die Schul?   Do you go to school?
Schwetscht du Deitsch?   Do you speak German?
Kannscht Deitsch   Can you speak German?
  schwetze?

3. This inverted word order is used also after interrogatives (question words):

Wu iss die Palli?   Where is Polly?
Waer bischt du?   Who are you?
Was iss dei Naame?   What is your name?

Note: As you can see from the E translations in 2. and 3., E follows the same rules of word order when questions begin with auxiliary verbs and interrogatives.

CHAPTER ONE

Part Two

Now let us try a more difficult reading selection,
using the same general vocabulary, but with more ad-
vanced word order and sentence structure. If you are
familiar with the vocabulary of the first reading
selection of this chapter, you should do very nicely.

Heit iss es saecktli zwansich Yaahr,
Dass ich bin owwenaus;
Nau bin ich widder lewich zrick
Un schteh am Schulhaus an der Grick,
Yuscht neegscht ans Daadis Haus.

Ich bin in hunnert Heiser gwest,
Vun Maerbelschtee un Brick,
Un alles, was sie hen, die Leit,
Deet ich verschwappe eenich Zeit
Fer's Schulhaus an der Grick.[2]

[2]*Henry Harbaugh, from "Das Alt Schulhaus an der*
Krick," Harbaugh's Harfe.

17

Der Aerscht Schuldaag; Es Zwett Deel

Die Schulbell ringt lang un laut, un die Schieler,
die Buwe un die Meed, laafe darrich der Schulhof un
gehne ins Schulhaus.

Es Schulhaus hot yuscht die ee Schtubb. Sie iss
net arrig grooss, awwer aa net zu glee fer die Schulkinner in der No(o)chberschaft.

Die Bell iss noch am Ringe, awwer deel Kinner hocke
schunnt uff de Benk un Schtiehl, die groosse links un
die gleene rechts. Der Schulmeeschder sitzt aa schunnt
am Schreibdisch. Er hot en Fedder in der Hand, guckt
uff en Schtick Babier--do sin viel Naame druff--un
schreibt alsemol ebbes nunner.

Dann un wann hebt er der Kopp un erkennt en Bu odder
en Meedel. Dann nuckt er un griesst sie mit, "Gut Marriye, Bill, wie bischt?" odder "Wie geht's demarriye,
Saelli?" Un sie andwadde, "Gut, danke, un du?"

Der Schulmeeschder weess, viele Kinner gleiche
Schul. Alle Daag kumme sie marriyeds viel zu frieh an
die Schul. Dann schtehne sie rum, renne rum, odder
schpiele Eckball uff em Schulhof, un waarde bis die
Bell ringt.

Nadierlich weess er aa, net yederebber gleicht
Schul. Deel gehne nie net uff Zeit noch der Schul un
sin alle Marriye schpot. Annere sin oftmols grank un
bleiwe deheem, un ebber iss immer am Schwense; er iss
immer am "Huckie" Schpiele. Awwer die menscht Zeit
sin sie yuscht in Zeit do un kumme in die Schulstubb,
wann sie heere, die Bell iss am Ringe.

Die Buwe un Meed hocke do un pischbere minanner
odder gucke rum. Der Meeschder schteht uff. "Dihr
wisst, Kinner, heit iss der aerscht Schuldaag. Do
hawwich die Naame vun de Kinner in der Klass. Deel
kennich schunnt, deel net. Nau, wu iss die Palli
Dreisbach?"

"Do binnich," saagt die Palli un schteht uff. Sie
iss schunnt elder, un sitzt links mit de annere
groosse Meed.

"Un duscht du Englisch schwetze?"

The First School Day; Part Two

    The school bell rings long and loudly, and the pupils, the boys and the girls, walk through the school yard and go into the schoolhouse.

    The schoolhouse has just the one room. It is not very large, but also not too small for the school children in the neighborhood.

    The bell is still ringing, but some children are already sitting on the benches and the chairs, the big (ones) left and the little (ones) right. The schoolmaster is also sitting already at the desk. He has a pen in his hand, looks down at a piece of paper--there are many names on it--and sometimes writes something down.

    Now and then he lifts his head and recognizes a boy or girl. Then he nods and greets them with, "Good morning, Bill, how are you?" or "How is it going this morning, Sally?" And they answer, "Good, thanks, and you?"

    The schoolmaster knows many children like school. Every day they come mornings much too early to school. Then they stand around, run around, or play corner ball in the schoolyard, and wait till the bell rings.

    Naturally, he knows also not everyone likes school. Some never go on time to school and are late every morning. Others are often sick and stay at home, and someone is always "bagging" school; he is always playing hooky. But most of the time they are here just on time and come into the schoolroom when they hear the bell is ringing.

    The boys and girls sit here and whisper together or look around. The master stands up. "You know, children, today is the first school day. Here I have the names of the children in the class. Some I know already, some not. Now, where is Polly Dreisbach?"

    "Here I am," says Polly and stands up. She is already older, and sits left with the other big girls.

    "And do you speak English?"

"Ya, gewiss, ich kann, un aa schpelle un schreiwe."

"Du kannscht Englisch schwetze. Sell iss gut. Du bischt schunnt en grooss Meedel, awwer du kannscht die Mudderschprooch schwetze. Sell iss aa gut."

Er guckt rum. "Tschanni Dreisbach, wu hockscht du denn?" Desmol saagt niemand nix. "Gebscht em Meeschder ken Andwatt?"

Dann saagt die Palli, "Seller iss mei Bruder. Er iss fer's aerscht mol in der Schul, un er verschteht dich net."

"Dutt dei Bruder net Englisch schwetze?"

"Nee, er dutt ken Watt schwetze. Awwer er iss net dumm. Er iss arrig gscheit un lannt schnell. Es nemmt net lang, schwetzt er gut Englisch."

Newwich em Tschanni hockt en Meedel. "Wie heescht du?" froogt der Meeschder.

"Ich bin die Lis Moyer, un do iss mei Schweschder. Sie heesst Till."

Un so geht's. Der Tietscher froogt, "Waer bischt du denn?" odder "Wie heescht du?" odder "Wu bischt du deheem?" un "Wu kummscht du haer?" odder "Wu bischt du denn haer?"

Deel kenne Englisch schwetze, annere kenne kee Englisch, yuscht Deitsch, schwetze; Englisch iss net so iesi. Deel wuhne uff re Bauerei uff em land, etliche wuhne im Schteddel. Deel kenne schunnt rechle, odder figgere, un kenne alle Buchschtaawe schee schreiwe. Sie kenne aa gut buchschtaawiere, odder schpelle, yuuse nie ken Schpellingbuch, un mache doch nie ken Fehler. Die Andwadde sin nie letz. Sie wisse alles.

Annere sin net abbaddich dumm, awwer sie loofe un fuule rum, kenne nie kee Frooge recht andwadde, un kenne nie nix im Kopp bhalde. Die Andwadde sin immer letz. Sie wisse nix.

Um zehe Uhr kann die Klass naus. Drauss kenne sie schpiele odder yuscht rumschteh un schwetze. Nau gucke mer rum. Was sehne mer do in der Schtubb? Nadierlich hot die Schtubb, wie alle Schtubbe, vier Wend, der Bodde(m) unne un die Deck owwe. Vanne iss es

"Yes, of course I can, and also spell and write."

"You can speak English. That is good. You are already a big girl, but you can speak the mother tongue. That is also good."

He looks around. "Johnny Dreisbach, where are you sitting?" This time no one says anything. "Don't you give the teacher an answer?"

Then Polly says, "That is my brother. He is in school for the first time, and he doesn't understand you."

"Doesn't your brother speak English?"

"No, he doesn't speak a word. But he is not stupid. He is very smart and learns quickly. It won't take long, he'll speak good English."

Next to Johnny sits a girl. "What is your name?" asks the teacher.

"I am Liz Moyer, and here is my sister. She is called Till."

And so it goes. The teacher asks, "Who are you?" or "What is your name?" or "Where do you live?" and "Where do you come from?" or "Where are you from?"

Some can speak English, others can speak no English, just German; English isn't easy. Some live on a farm in the country, some live in town. Some can already do arithmetic, and can write all letters nicely. They can also spell well, never use a spelling book, and nevertheless make no mistakes. The answers are never wrong. They know everything.

Others are not especially stupid, but they loaf and fool around, can answer no question correctly, and can never keep anything in their heads. The answers are always wrong. They know nothing.

At ten o'clock the class can (go) out. Outside they can play or just stand around and talk. Now we'll take a look around. What do we see here in the room? Naturally the room has, like all rooms, four walls, the floor below and the ceiling above. In front is the

Schwazbord un aa en Dier. Sie iss zu. Iwwer em Blaeckbord hengt die grooss Uhr.

Links hot die Wand drei Fenschdere. Die Wand rechts hot yuscht zwee Fenschdere--eens iss schunnt uff--awwer sie hot aa en Dier. Die Wand hinne hot ken Fenschder un ken Dier. Die Bilder un die Buchschtaawe henge datt. Im Winder schteht aa der Offe in eenre Eck vun der Schtubb.

Der Tietscher schreibt uffs Blaeckbord mit Greid. Nadierlich yuust er aa Fedder un Dinde. Die groosse Kinner schreiwe aa mit Dinde, awwer die gleene kenne en Bleibensil yuuse. Es Schulkind kann uff Babier odder in en Kappibuch schreiwe. Uffkoors hot yeder-ebber sei Schpellingbuch un lest alle Daag aa annere Bicher.

Vocabulary

    abbad'dich: especially
    aensere: to answer
    allgebott: now and then
    alsemol: sometimes, at times
    andwadde: to answer
die Bauerei', Bauerei'e: farm
    bhalde: to keep
 es Bild, Bilder: picture
    bleiwe: to remain
der Bu, Buwe: boy
der Buchschtaab,-schtaawe: letter of alphabet
    buchschtaawier'e: to spell
    danke: thanks, thank you
    dann un wann: now and then
    darrich: through
    deheem': at home
    demar'riye: this morning
    denn: (see vocabulary note)
    desmol: this time
    druff: on it, upon it
    duh: to do
    ebber: someone
    ebbes: something
die Eck, Ecke: corner; Eckball: corner ball
    ee: one

blackboard and also a door. It is closed. Over the blackboard hangs the large clock.

To the left the wall has three windows. The wall to the right has just two windows--one is already open-- but it also has a door. The wall in back has no window and no door. Pictures and the letters hang there. In winter the stove also stands in one corner of the room.

The teacher writes on the blackboard with chalk. Naturally he uses pen and ink too. The big children also write with ink, but the little (ones) can use a lead pencil. The school child can write on paper or in a copy book. Of course, everyone has his spelling book and reads every day other books too.

    elder: older
    erken'ne: to recognize
    etliche: some, a few
der Fehler, Fehler: mistake
    fer: for
    figgere: to figure (do arithmetic)
    frieh: early
die Froog, Frooge: question
    fuule: to fool
    gewwe: to give
    gleiche: to like
    grank: sick
    griesse: to greet
    gucke: to look, peer
die Hand, Hend: hand
    heere: to hear
    hewe: to lift
    hinne: in back, in the rear
    iesi: easy
    immer: always
    iwwer: over
 es Kappibuch,-bicher: copy book
    ken, kee: no, not a, none
der Kopp, Kepp: head
    laafe: to walk
    lang: long
    laut: loud, loudly
    letz: wrong

loofe: to loaf
mache: to make, to do
marriyeds: mornings, in the morning
es Meedel, Meed: girl
menscht: most
minan'ner (mitnan'ner): together
mol(l): once, at some time
die Mudderschprooch: mother tongue
nadier'lich: naturally, of course
nau: now
naus: out
nei: new
nemme: to take
newwich: beside, next to
niemand: no one, nobody
noch: still, yet
noch: to
die Nochberschaft: neighborhood
nucke: to nod
nunner: down
der Offe, Effe: stove
oftmols: often
pischbere: to whisper
rechle: to reckon (do arithmetic)
recht: right, correct
renne: to run
rum: around
schee: nice, beautiful
schpelle: to spell
schpiele: to play
schpot: late
die Schprooch, Schprooche: language
es Schtick, Schticker: piece; en Schtick Babier: a
    piece of paper
der Schuler, Schieler (or Schuler): pupil
der Schulhof,-heef: schoolyard
schwense: to play hooky, "bag" school
die Schweschder, Schweschdere: sister
sell: that
so: so, such
uff: up
uffkoors': of course
vun: of, from
waarde: to wait

der Winder, Windere: winter
   yederebber: everyone
   zamme: together
   zehe: ten
die Zeit, Zeide: time
   zu: to, too
   zu: closed

Vocabulary Notes

figgere, fuule, gleiche, laafe, loofe, mache, nemme, nucke, pischbere, renne, schpelle, and schpiele are conjugated like kumme.

buchschtaawiere is one of several PG verbs we can refer to as -iere verbs. They are always accented on the syllable -ier.

bleiwe, gewwe, and hewe are conjugated like schreiwe.

griesse and schwense are conjugated like schwetze.

der Buchschtaab, Buchschtaawe: Note that a final b changes to w before an ending that is a vowel or begins with a vowel.

die Zeit, Zeide: Note that a final t changes to d before an ending that is a vowel or begins with a vowel.

die Froog, Frooge: In keeping with the rules of spelling and pronunciation, the plural of Froog is spelled Frooge, and pronounced by many (not necessarily all) PG speakers as if it were spelled Froo-e.

figgere: This verb is spelled with gg to show that the preceding vowel i is short. The verb is pronounced much like fi-ere by most (not necessarily all) PG speakers.

denn: This word is a coordinating conjunction that introduces main clauses, but it is also a particle used very often in PG questions, and if you get into the habit of using it too, your questions will sound much more natural. Only rarely can a good translation of this particle be pinpointed. It might imply surprise or impatience, but usually an E equivalent is rendered not by a specific word, but by the tone of voice.

erkenne: This verb is conjugated exactly like kenne
"to know (a fact)." Er- is another of the four common prefixes that never separate from the verb and
are never accented.

es Bild: This German word is often replaced with the
very popular E loan word es Pikter, Pikters.

niemand: also nimman(d).

Grammar

A. The Definite Article, Dative Case

1. As we have seen, the PG definite articles differ
from gender to gender and from number to number. We
have also noted that the nominative and accusative case
forms are the same for each gender. The dative
articles, however, differ from this pattern.

	Singular			Plural
	M.	F.	N.	All Genders
	em	der	em	de

Here now is the entire declension:

	Singular			Plural
	M.	F.	N.	All Genders
N. and A.	der	die	es	die
Dative	em	der	em	de

2. The dative case is the case of the indirect object, as distinguished from the direct object.

Ich	geb	em Meeschder	es Buch.
s.	v.	indirect obj.	dir. obj.
I	give	(to) the teacher	the book.
Du	gebscht	der Meeschdern(f.)	es Buch.
Mer	gewwe	de Kinner	die Bicher.

3. A good rule of thumb for a verb such as gewwe:
the direct object is whatever you give, and the indirect object is whomever you give it to. There are
other such verbs, of course, and they will be pointed
out as they appear in subsequent vocabulary lists.

4. The dative case, as well as the accusative case, must be used after various prepositions. Some examples taken from the preceding reading selections follow:

Accusative:
darrich der Schulhof   : through the schoolyard
in die Schulschtubb   : into the school room
ins (in+es) Schulhaus   : into the school house

Dative:
uff em Schulhof   : on the schoolyard
in der Klass   : in the room
vun de Kinner   : of the children

We will be studying prepositions in detail in Chapter 6. Until then, simply accept the prepositional phrases as you find them in the reading selections, much as you would if you were living in a PG household and picking up the language in day-to-day conversation.

5. Other uses of the dative case will be pointed out as necessary in following chapters.

B. The Indefinite Article, Dative Case

1. We have also noted that the indefinite article is en for all genders in both the nominative and accusative cases. Again, the dative case breaks this pattern.

	Singular		Plural
M.	F.	N.	All Genders
me	re	me	No form

Thus the following declension:

	Singular		
	M.	F.	N.
N. and A.	en	en	en
D.	me	re	me

2. The following sentences illustrate the use of the indefinite articles in the nominative, accusative, and dative cases:

En Bu gebt me Tietscher A boy gives a teacher a
en Buch. book.
En Meedel gebt re A girl gives a teacher(f.)
Tietschern en Fedder. a pen.

C. <u>ken</u>, <u>kee</u>: "no," "none"

1. Because the indefinite article has no plural
forms, let us go immediately to <u>ken</u> and <u>kee</u>. Just as
E speakers do not normally say, "I have not a book,"
but rather "I have no book," the PG speaker does not
normally say <u>Ich hab net en Buch</u>, but <u>Ich hab ken (kee)</u>
<u>Buch</u>, or in the plural, <u>Ich hab ken (kee) Bicher</u>. Ken
and <u>kee</u> are "contractions" of <u>net</u> and an indefinite
article. But note that this negative has a plural
form also:

	Singular			Plural
	M.	F.	N.	All Genders
N. & A.	ken,kee	ken,kee	ken,kee	ken,kee
D.	kem	kenre	kem	ken

2. Younger speakers appear to prefer <u>ken</u> to <u>kee</u>,
perhaps because it so closely resembles the pattern of
the indefinite article <u>en</u>.

3. What has been said about definite and indefinite
articles following verbs and prepositions can also be
said about <u>ken</u>:

Er schwetzt ken Watt. He speaks no word.
Mer gewwe kem Kind ken We give no child a pencil.
Bensil.

D. Verbs

1. <u>duh</u> "to do": This verb is conjugated very much
like <u>geh</u> and <u>schteh</u>, but it is different enough to present it in its entirety:

Singular	Plural
1. ich duh	1. mir,mer duhne
2. du duscht	2. dihr dutt, ihr duhne, etc.
3. er,sie,es dutt	3. sie duhne

(Duhn is a variation in the plurals: mir, dihr, sie
duhn.)

2. But duh is like geh and schteh in that all three
add -n whenever they are used with inverted word order:
gehn ich, schtehn ich, duhn ich.   Actually, the verb
and the pronoun usually sound like one word, e.g.,
duhnich.

3. PG speakers use duh very much as E speakers use
the emphatic "do," though usually only to ask questions or to make negative statements.   But whatever
its use, it resembles kenne "to be able to" in the
following respects:

Kannscht du Deitsch schwetze?	Can you speak German?
Duscht Deitsch schwetze?	Do you speak German?
Ich kann net Deitsch schwetze.	I cannot speak German.
Ich duh net Deitsch schwetze.	I do not speak German.

4. The PG verb rechle "to figure (do arithmetic)"
is an example of a group of verbs whose infinitives
end in a consonant + le.   It has the following forms:

Singular	Plural
1. ich rechel	1. mir,mer rechle
2. du rechelscht	2. dihr,der rechelt; ihr rechle, etc.
3. er,sie,es rechelt	3. sie rechle

Other verbs in this group will be pointed out as
they appear in the vocabulary lists.

By the way, schpiele and schpelle are not two of
these verbs because their endings -le and -lle are
preceded by vowels, not by consonants.

5. For the sake of simplicity, verbs whose stems
end in d will be conjugated like bhalde "to keep":

Singular	Plural
1. ich bhald	1. mir,mer bhalde
2. du bhaldscht	2. dihr,der bhaldt; ihr,er bhalde; etc.
3. er,sie,es bhaldt	3. sie bhalde

6. But the conjugation of verbs whose stems end in <u>dd</u> will follow the B/B system more closely. <u>Andwadde</u> "to answer" is an example from this chapter:

Singular	Plural
1. ich andwatt	1. mir,mer andwadde
2. du andwattscht	2. dihr,der andwatt; ihr,er andwadde; etc.
3. er,sie,es andwatt	3. sie andwadde

E. Word Order

1. The first reading selection introduced us to normal word order as well as inverted word order for questions:

Er   schwetzt   Deitsch.

Schwetzt   er Deitsch?

Was   schwetzt   er?

We learned that the verb is normally the second element in a declarative sentence (clause). PG speakers also use inverted word order when they wish to emphasize an element of the sentence other than the subject, very much as an E speaker changes from "The clock hangs over the blackboard" to "Over the blackboard hangs the clock," (rather than "Over the blackboard the clock hangs").

Die Uhr   hengt   iwwer em Blaeckbord.

Iwwer em Blaeckbord   hengt   die Uhr.

Other examples of this use of inverted word order taken from the preceding text:

Vanne iss es Schwazbord.	In front is the blackboard.
Um zehe Uhr kann die Klass naus.	At ten o'clock the class can go out.

2. But only <u>one</u> element (and its modifiers) is allowed to precede the verb:

Der Offe	schteht	im Winder in der Eck.
Im Winder	schteht	der Offe in der Eck.
In der Eck	schteht	der Offe im Winder.

You will probably never hear a PG speaker say, <u>Im</u> <u>Winder in der Eck schteht der Offe</u>. Also, as the first of the above three sentences shows, he prefers to place adverbs of time before adverbs of place, and so likes to begin sentences with time expressions, as in the second of the three examples above. The word order of the third sentence must be considered relatively rare.

F. Double Negatives

1. Our English teachers tell us that two negatives make a positive; if we say, "We never have no money," we are in reality saying that we always have money. But some of us don't pay any attention to them and believe that the double negative serves to strengthen our negative statements. So do the PG speakers, who strongly believe that a double negative leaves no doubt as to the negative meaning: <u>Ich yuus nie ken</u> <u>Schpellingbuch, un ich mach nie ken Mischtaiks</u> "I never use no spelling book, and I never make no mistakes."

G. The PG "Progressive"

1. Normally, the one PG form <u>ich schreib</u> can be the equivalent of the three E forms "I write," "I do write," "I am writing." We have already seen that under certain circumstances, <u>duh</u> is used by PG speakers much as the emphatic auxiliary "do" is used in English. But PG speakers have no literal equivalent of the E progressive, for example "I am writing, you are writing," and the like. But they do have a construction with which they can express the progressive aspect of the verb: <u>sei</u> + <u>am</u> + infinitive. The present tense of <u>sei</u> is used to render the present progressive, e.g.

Ich bin am Schreiwe.   I am writing.
Er iss am Lese.   He is reading.

Note the capitals! The infinitives are actually nouns, the objects of the preposition <u>am</u> (<u>an</u>+<u>em</u> = <u>am</u>).

Part One

The vocabulary of the family is certainly close to
all of us. We probably use it more consistently
through our lives than any other vocabulary. We
will begin with a simple reading selection using only
the basic vocabulary of the family, the words we can
start to use at home this very day.

Liewi Mudder, du bischt mir neegscht
Wann du dei Hand in meini leegscht
  Un blauderscht weil mit mir;
Dei Schprooch iss doch so siess un sacht--
Noh guck ich rum un nemm in acht
  'S iss nur en Draam vun dir.[1]

Der Daadi lest die alde Biwwel;
Er hasst die Sind un alles Iwwel.
Im Schtille er oft hazlich beet,
Dass Gott sie all erhalde deet.[2]

Die Eldre fiehle schtolls un froh--
Sie hen en Bobbli--'s iss en Soh.
Die Nuus geht rum, un zimmlich glei
Viel Freind un Nochbere kumme bei.[3]

[1] *Ralph Funk, from "Zum Andenke,"* <u>PG Verse</u>.

[2] *Henry Moyer, from "Der Alde Schannschtee,"* <u>PG Verse</u>.

[3] *Charles C. Ziegler, from "Die Eldre Fiehle Schtolz*
  un Froh," <u>PG Verse</u>.

Die Familye; Es Aerscht Deel

Viel deitsche Familye sin wunnerbaar grooss. Awwer
mei Familye iss net so grooss. Ich hab nadierlich en
Vadder un en Mudder; sie sin mei Eldre. Mei Vadder
iss meinre Mudder ihre Mann, un mei Mudder iss sei
Fraa.

Ich bin meine Eldre ihre Suh, awwer sie hen zwee
meh Seh; sie sin mei Brieder. Un mei Eldre hen drei
Dechder; sie sin mei Schweschdere un ich bin ihre Bruder.

Meim Daadi sei Familye iss aa net grooss. Sei
Daadi, mei Groossdaadi, lebt noch. Mei Groossmammi,
meim Paepp sei Mudder, iss aa noch am Lewe. Ich bin
ihre Kinskind, odder Enkel, un mei Memm iss ihre
Suhnsfraa. Mei Groosseldre sin meinre Memm ihre
Schwaermudder un Schwaerdaadi.

Mei Vadder hot zwee Brieder, mei Onkel (Unkel); un
ee Schweschder, mei Ant (Aent). Ich bin ihre Bruderskind. Ee Bruder iss leddich.

Meinre Memm ihre Familye iss glee. Sie hot yuscht
ee Bruder un ee Schweschder. Ich bin ihre Schweschderskind. Ihre Eldre lewe nimmi; sie sin schunn lang
dot.

Mei eldschder Bruder iss verheiert un wuhnt nimmi
bei uns. Aer un sei Fraa hen en Wohning in der
Schtadt. Sie hen en Beewi, en gleener Bu. Ich bsuch
sie oft.

Mei zwett eldschdi Schweschder hot en Bo. Verleicht heiert sie en ball. Dann hen sie en Hochzich
un gehne uff die Hochzichrees. Un dann kann ich widder en Onkel warre (waerre).

Vocabulary

die Aent, Aents: aunt
die Ant, Ants: aunt
   ball: soon
es Beewi, Beewis: baby
   bei: at, with

The Family; Part One

Many German families are "wonderfully" large. But
my family is not so large. I have naturally a father
and a mother; they are my parents. My father is my
mother's husband, and my mother is his wife.

I am my parents' son, but they have two more sons;
they are my brothers. And my parents have three
daughters; they are my sisters and I am their
brother.

My father's family is also not large. His father
(daddy), my grandfather, is still living. My Grandma, my father's mother, is also still living. I am
their grandchild, and my mom is their daughter-in-law
(son's wife). My grandparents are my mom's mother-
and father-in-law.

My father has two brothers, my uncles; and one
sister, my aunt. I am their nephew (brother's child).
One brother is single (unmarried).

My mother's family is small. She has just one
brother and one sister. I am their nephew (sister's
child). Her parents aren't living any more; they are
(have been) dead a long time.

My oldest brother is married and lives no longer
with us. He and his wife have a residence in the
city. They have a baby, a little boy. I visit them
often.

My second oldest sister has a beau. Perhaps she
will marry him soon. Then they will have a wedding
and go on the wedding trip. And then I can become an
uncle again.

der Bo, Bos: beau, boyfriend
 es Bruderskind, -kinner: nephew, niece
der Daadi, Daadis: daddy, father
die Dochder, Dechder: daughter
der Dochdersmann, -menner: son-in-law
    dot: dead
die Eldre (pl.): parents
    eldscht: oldest

der Enkel: grandson
die Famil'ye: family
die Fraa, Weiwer: wife, woman
der Groossdaadi, -daadis: grandfather
die Groosseldre: grandparents
die Groossmammi, -mammis: grandmother
    hawwe: to have
    heiere: to marry
die Hochzich: wedding
die Hochzichrees: wedding trip
 es Kinskind, -kinner: grandchild
    lewe: to live
die Mammi, -s: mommy, mother
der Mann, Menner: man, husband
    meh: more
die Memm, -s: mom, mother
die Mudder, Midder: mother
    nimmi: no longer
    noch: yet, still
der Onkel (Unkel): uncle
der Paepp: father, "pop," "pap"
die Schtadt, Schtedt (Schtadde): city
der Schwaerdaadi, -s: father-in-law
die Schwaermudder, -s: mother-in-law
 es Schweschderskind, -kinner: nephew, niece
der Suh, Seh: son
die Suhnsfraa, -weiwer: daughter-in-law
der Vadder, Vedder: father
    verhei'ert: married
    verleicht': perhaps, maybe
    warre (waerre): to become
    widder: again
die Wohning, Wohninge: dwelling, residence
    wunnerbaar: wonderful(ly)
    yuscht: just
    zimmlich: quite, rather
    zwee: two
    zwett: second

Vocabulary Notes

hei(e)re "to marry" is conjugated like kumme.

lewe "to live" is conjugated like schreiwe.

bei: This preposition is used as an equivalent of E "with" or "at" to express the idea of living "with" or "at the home" of a person or family. Mei Bruder wuhnt bei meim Onkel "my brother lives with my uncle (at the home of my uncle)." Merely to express the idea "at the Dreisbachs," as in, for instance "I was just at the Dreisbachs," the PG speaker uses the preposition ans and adds an -s to the name: Ich waar yuscht ans Dreisbachs.

die Dochder also has the plural form Dochdere.

die Eldre: This noun appears in the plural only; there is no PG equivalent of the singular E noun "parent."

die Fraa: This noun is used much more to mean "wife" than to mean "woman," which is usually expressed by the noun es Weibsmensch, Weibsleit.

die Hochzichrees has the variant die Hochzichdripp, in which -dripp is the E loan word "trip."

der Schwaerdaadi also pronounced and written Schwier- and Schwar-.

der Suh, Seh: also Soh, Seh.

verleicht: also villeicht.

Grammar

A. Personal Pronouns, All Cases

1. It is time to expand our knowledge of the personal pronouns. Unlike the definite and indefinite articles, almost all personal pronouns have different forms in the nominative and accusative cases.

		Singular					Plural		
		Acc.	Unacc.				Acc.	Unacc.	
1.	N.	ich	ich	I	1.	N.	mir	mer	we
	A.	mich	mich	me		A.	uns	uns	us
	D.	mir	mer	to me		D.	uns	uns	to us
2.	N.	du	d(e)	you	2.	N.	dihr	der	
							ihr	er	
							nihr	ner	you
	A.	dich	dich	you		A.	eich	eich	you
	D.	dir	der	to you		D.	eich	eich	to you

	Singular Acc. Unacc.				Plural Acc. Unacc.		
3. Masculine							
N. aer	er	he					
A. ihn	en	him					
D. ihm	em	to him					
Feminine							
N. sie	se	she		3. N. sie	sie	they	
A. sie	se	her		A. sie	sie	them	
D. ihre	re	to her		D. ihne	ne	to them	
Neuter							
N. es	's	it					
A. es	's	it					
D. ihm	em	to it					

B. Possessive Adjectives

1. Every personal pronoun has a corresponding possessive adjective and possessive pronoun form. The possessive adjectives are normally placed before a noun (or another noun modifier), e.g., my book, your new car, their old and worn out car.

	Singular				Plural		
1. ich	mei	my		1. mir,mer	unser	our	
2. du	dei	your		2. dihr,ihr, nihr	eier	your	
3. M. er	sei	his		3. sie	ihre	their	
F. sie	ihre	her					
N. es	sei	its					

2. But just as the personal pronouns have various forms for the various cases, so do the possessive adjectives:

a) mei "my"

	Singular			Plural
	M.	F.	N.	All Genders
N. & A.	mei	mei	mei	mei
D.	meim	meinre	meim	meine

This paradigm may serve also as a model for dei "your" and sei "his," "its."

b) <u>ihre</u> "her," "their"

	Singular			Plural
	M.	F.	N.	All Genders
N. & A.	ihre	ihre	ihre	ihre
D.	ihrem	ihrer	ihrem	ihre

c) <u>unser</u> "our" and <u>eier</u> "your"

Singular

	M.	F.	N.
N.& A.	unser,eier	unser,eier	unser,eier
D.	unserm,eierm	unserer,eierer	unserm,eierm

Plural

All Genders

N. & A.	unser,eier
D.	unsere,eiere

3. Many foreign language students have difficulty understanding three concepts concerning possessives. The first is the difficulty of reconciling the term possessive (genetive) and the fact that possessives can appear in the nominative, accusative, or dative cases. The second concerns the fact that a masculine possessive can modify a feminine or neuter noun, a feminine possessive can modify a masculine or neuter noun, and a neuter possessive can modify a masculine or a feminine noun. The third concerns the ability of singular possessives to modify plural nouns and plural possessives to modify singular nouns. Hopefully the following sentences will eliminate these difficulties.

Sei Schweschder	kennt		ihre Bruder.
subject	v.		direct obj.
nominative			accusative
His sister	knows		her brother.

Em Kind sei Eldre	sin	aa	ihre Verwande.
subject	v.		predicate noun
nominative			nominative
The child's parents	are	also	her relatives.

Ihre Vadder	bsucht		unser Mudder.
subject	v.		direct object
nominative			accusative
Their father	visits		our mother.

C. Possessive Pronouns

1. Possessives can also be pronouns; that is, they can take the place of a noun. <u>Meiner</u> "mine" can serve as a model for all possessives used as pronouns.

	Singular			Plural
	M.	F.	N.	All Genders
N.& A.	meiner	meini	meins	meini
D.	meim	meinre	meim	meine

2. The following sentences illustrate the use of possessive adjectives and pronouns:

Ihre Bruder iss grooss, awwer meiner iss glee.	Her brother is big, but mine is small.
Unser Familye iss grooss; iss eieri aa grooss?	Our family is large; is yours also large?
Du hoscht dei Buch un ich hab meins.	You have your book and I have mine.
Ich geb deinre Mudder mei Bensil; gebscht du meinre deins?	I give your mother my pencil; do you give mine yours? (or do you give yours to mine?)

D. Use of the Dative Case to Show Possession

1. We see, then, that E expressions such as "my books," "your house," "their brother" have literal (word for word) equivalents in PG: <u>mei Bicher</u>, <u>dei Haus</u>, <u>ihre Bruder</u>. There are, however, no literal equivalents for the following E possessive expressions: my daddy's books, the parents' house, that mother's sister. These E expressions must be rendered by means of the PG combination of the possessor in the dative case, followed by the possessive adjective, followed by the thing or person possessed. As always, no matter what the circumstances, the possessive must reflect the gender and number of the antecedent (the noun whose place the pronoun takes), but the case of the possessive is determined by its use in the phrase or clause in which it appears.

article or adjective in dative case	+ possessor	+ poss. adj.	thing or object + possessed
meim (to) my	Daadi daddy	sei his	Bicher books
de (to) the	Eldre parents	ihre their	Haus house
daere (to) that	Mudder mother	ihre her	Schweschder sister

The idiomatic translations of the above PG expressions would be "my father's (daddy's) books," "the parents' house," and "that mother's sister."

Some such expressions from the preceding reading selection follow:

meim Daadi sei Familye	my father's family
meim Paepp sei Mudder	my dad's (pop's) mother
meinre Memm ihre Schwaer-mudder	my mother's mother-in-law

E. Verbs

1. hawwe "to have"

We have been using various forms of the verb hawwe "to have" since the very first lesson. Here, now, is the entire conjugation:

Singular	Plural
1. ich hab	1. mir,mer hen
2. du hoscht	2. dihr,der; ihr,er; nihr,ner hen
	dihr,der; ihr,er hett
	dihr,der hend
3. er,sie,es hot	3. sie hen

2. When this verb is used in inverted word order, the following forms are used more closely to reflect what happens in the spoken language:

Ich hab en Buch.	I have a book.
Hawwich en Buch?	Do I have a book?

Mer hen en Bensil.   We have a pencil.
Hemmer en Bensil?   Do we have a pencil?

Ihr hen en Fedder.   You have a pen.
Henner en Fedder?   Do you have a pen?

Hodder (hot er) is rarely found in PG literature, but it is a close approximation of the spoken form.

The second person singular du is often "swallowed" if it follows the verb:

Du hoscht en Buch.   You have a book.
Hoscht en Buch?   Do you have a book?

Some PG speakers use a very weak, unaccented de: Hoscht de en Buch? This form could possibly be represented by hoschde or hosch de, but these representations are rare in the written language.

Starting with the next lesson, the second person singular du will be deleted in inverted word order: Hoscht en Buch? Gehscht ins Schteddel? Kummscht net heit?

But accented pronoun du will not be deleted:

Ich hab en Buch. Hoscht   I have a book; do you
du aa eens?   have one too?

3. warre (waerre) "to become"

Warre (or waerre) is the third of the "big three"-- sei, hawwe, and warre--that should be practiced until you know them forwards and backwards. Not only are they important in their own right, but they will be used as helping verbs (auxiliaries) very soon in future lessons.

Singular	Plural
1. ich wa	1. mir,mer warre
2. du wascht	2. dihr,der; ihr,er; nihr,ner warre
	dihr,der; ihr,er watt
	dihr,der warret
3. er,sie,es watt	3. sie warre

CHAPTER TWO

Part Two

This second reading selection is much more involved
and contains a wider range of vocabulary than the
first. Perhaps you want to use it at first only as a
reading practice to increase your passive vocabulary.
But eventually you will want to make all of the expressions and grammar a part of your active knowledge
of PG.

Ich hab en liewi Widfraa gheiert,
Die hot en g'wachseni Dochder;
Mei Vadder waar en Widmann gwest
Un heiert mei Schtiefdochder;
So waar mei Daed mei Dochdermann,
Un ich sei Schwiegervadder;
Sei Fraa, die waar mei Dochder gwest,
Un allerdings mei Mudder;
Wie ich en Zeitlang gheiert waar,
Do grickt mei Fraa en Suh.
Der waar dann Schwoger zu meim Vadder,
Un ich waar sei Neffyu;
Meim Daed sei Fraa grickt aa en Suh,
Der waar, uffkoors, mei Bruder--
Zu gleiche Zeit mei Enkelche,
Dann, ich waar sei Groossvadder,
Mei Fraa, die waar mei Gremmemm gwest,
Un ich waar ihre Enkel;
Ich waar mei eegner Graendaed, 's letscht
Un dann waar's Zeit zu henke.[1]

[1] *Henry L. Fischer, "En Verwickelte Verwandschaft,"*
PG Verse.

43

Die Familye; Es Zwett Deel

Viele deitsche Familye sin wunnerbaar grooss--zehe, elfe, zwelfe un meh. Awwer mei Familye iss net abbaddich grooss; mer sin zamme yuscht neine. Nadierlich hawwich en Vadder un en Mudder.

Ich hab aa noch drei Brieder un des meent mei Eldre hen vier Buwe, odder Seh. Ich hab aa drei Schweschdere; die sin meine Eldre ihre Meed, odder Dechder, un ich bin ihre Bruder. Des meent ich hab sex G(e)schwischder.

Mer wuhne all uff re Bauerei uff em Land. Der Daadi schafft hatt vun frieh bis schpot uff der Bauerei, awwer mei Brieder un ich helfe ihm, wann mer kenne. Mei Mudder, en geborni Yoder, schafft aa abbaddich hatt. Ihre Leit sin all gude Bauersleit, un wuhne immer noch uff de beschte Bauereie in daere Kaundi. Sie iss immer am Butze odder am Koche, awwer mei eldschdi Schweschder helft re alle Daag.

Meim Daadi sei Familye iss aa net grooss, yuscht siwwe. Er hot vier Gschwischder, drei Brieder--zwee sin Zwillinge--un ee Schweschder. Die iss awwer schunn Witfraa. Em Paepp sei Brieder sin mei Onkel, un sei Schweschder iss mei Ant.

Ich gleich sie all; die sin all in der Freindschaft un sin gude Verwande. Die sin immer gut zu mir. Ich bsuch sie oft, un dann gewwe sie mer Kaendi un alsemol fimf odder zehe Zent. Des gleich ich aa.

Meim Paepp sei Paepp, mei Groossdaadi, lebt noch. Ich denk er iss arrig alt; er iss blod, baalkeppich, un er hot en langer Baart. Awwer er iss immer frieh uff un schafft enwennich alle Daag. Ich gleich en. Ya, ich lieb mei Gremmpaepp sogaar.

Mei Groossmemm, meim Paepp sei Mudder, iss aa noch am Lewe. Ich sehn sie iss am Altwarre. Sie iss grummbucklich un runslich, un ihre Haar iss weiss wie Schnee. Sie sitzt gaern im Schockelschtuhl un schockelt, awwer sie kann noch wunnerbaar gut koche un Kuche un Boi backe. Ich gleich ihre Koches un Backes.

Uff meinre Memm ihre Seit vun der Freindschaft hawwich yuscht ee Onkel un ee Aent. Ihre Eldre lewe nimmi; sie sin schunn lang dot un leige im Karrichhof.

The Family; Part Two

Many German families are wonderfully large--ten, eleven, twelve and more. But my family is not especially large; together we are only nine. Naturally I have a father and a mother.

I have also three more brothers and that means my parents have four boys, or sons. I also have three sisters; they are my parents' girls, or daughters, and I am their brother. That means I have six siblings.

We all live on a farm in the country. Dad works hard from early to late on the farm, but my brothers and I help him when we can. My mother, nee Yoder (maiden name Yoder), also works especially hard. Her people are all good farm people, and still live on the best farms in this county. She is always cleaning or cooking, but my oldest sister helps her every day.

My dad's family is also not large, just seven. He has four siblings, three brothers--two are twins--and one sister. But she is already a widow. Dad's brothers are my uncles, and his sister is my aunt.

I like them all; they are all in the family (in the largest sense) and are good relatives. They are always good to me. I visit them often and then they give me candy and sometimes five or ten cents. I like this too.

My dad's dad, my grandfather, is still living. I think he is very old; he is bald, bald-headed, and he has a long beard. But he is always up early and works a little every day. I like him. Yes, I love my granddad even.

My grandma, my father's mother, is also still living. I see she is getting old (becoming old). She is bent over (humpbacked) and wrinkled, and her hair is white as snow. She likes to sit in the rocking chair and rock, but she can still cook wonderfully and bake cakes and pie. I like her cooking and baking.

On my mother's side of the family I have just one uncle and one aunt. Her parents are no longer alive; they have already been dead a long time and are lying in the churchyard.

Mei eldschder Bruder iss schunnt gheiert. Die iss awwer en Weibsmensch! Gaar net wiescht, gutguckich sogaar, awwer was en Babbelmaul, saag ich eich. Un ihre Leit sin Schtadtleit. So geht's.

Sie hen en Bobbel; des iss mei Bruderskind. Un ich bin nadierlich sei Onkel. Ball watt er elder, dann kann ich em dann un wann fimf Zent gewwe. Ya, ich lieb der glee Schpringer. Awwer zehe Zent kann zu deier warre.

Mei Groossdaadi iss sei Urgroossdaadi, mei Groossmammi iss sei Urgroossmammi, un aer iss ihre Urenkel.

Mei eldschdi Schweschder is verschproche. Ich glaab, er iss kenner vun denne gude ehrliche Kalls vun kennre Bauersfamilye. Er schafft in re Faeckdri odder uff em Riggelweg, glaawich. Was weess ich. Un geizich? Ya, ich kenn en gut. Ich saag eich, kenner iss so en Geizhals. Nee, daer iss ken Tschentelmann. Un ich drau em net. Nee, dem drau ich net. Er gebt mer nie nix, neddemol en Zent.

Awwer die zwee heiere, des weess ich. Dann ziehge sie noch der Schtadt. So geht's.

Mei zwett eldschdi Schweschder hot en Bo. Sie schpringe schunn lang zamme. Was en Mannskall! En gudi Maetsch, denk ich. Er iss immer schwer am Schaffe--helft sogaar meim Vadder--verdient net viel, awwer genunk. Er gebt mer fimfunzwansich Zent.

Ya, ich gleich mei Schweschder un ihre Bo. Un wann sie heiere, dann geh ich mit denne zwee uff die Hochzichrees. Glaawich.

Ennihau, des nemmt dann net lang, do hen sie en Beewi, un do wa ich widder Onkel. Ich weess, wie des geht.

Vocabulary

alt: old; am Altwarre: getting old
baalkeppich: bald-headed
der Baart, Beert: beard
es Babbelmaul,-meiler: blabbermouth, chatterbox
backe: to bake

My oldest brother is already married. My, but
she's some woman! Not at all ugly, even good looking,
but what a blabbermouth, I tell you. And her people
(family) are city people. That's the way it goes.

They have a baby; that's my nephew. And I am
naturally his uncle. Soon he will get older; then I
can give him five cents now and then. Yes, I love
that little fellow. But ten cents can get too expensive.

My grandfather is his great grandfather, my grandma is his great grandma, and he is their great grandson.

My oldest sister is "promised." I believe he is
not one of those good, honest fellows from a farm family. He works in a factory or on the railroad, I believe. What do I know. And stingy? Yes, I know him
well. I tell you, no one is such a miser. No, he is
no gentleman. And I don't trust him. No, him I don't
trust. He never gives me anything, not even a cent.

But those two will marry; that I know. Then they
will move to the city. That's the way it goes.

My second oldest sister has a beau. They have been
going together for a long time. What a man! A good
match, I think. He is always working hard--even helps
my father--doesn't earn much, but enough. He gives me
twenty-five cents.

Yes, I like my sister and her boyfriend. And when
they marry, then I'll go on the wedding trip with
those two. I believe.

Anyhow, that won't take long, then they will have a
baby, and then I will become an uncle again. I know
how that goes.

 es Backes: baking
die Bauersfamilye: farm family
die Bauereleit: farmers, farm people
    bescht: best
    blod: bald
 es Bobbel, Bobbelcher: baby
der Boi, Bois: pie

```
    deier:  expensive
    denke:  to think
    ehrlich:  honest
    elf(e):  eleven
    ennihau:  anyhow
    enwen'nich:  a little
die Faeckdri, -s:  factory
    fimf:  five
    fimfunzwansich:  twenty-five
die Freindschaft:  relatives, relation
    gaar ken:  none at all
    gaar net:  not at all
    gaern:  like (see note)
    gebor'ni:  nee
der Geizhals:  stingy person
die Geschwisch'der:  siblings
    gewwe:  to give
    gheirt:  married
    glaawe:  to believe
der Gremmpaepp:  grandfather
die Groossmemm:  grandmother
    grummbucklich:  humpbacked
    gutguckich:  good looking
 es Haar, Haare:  hair
    hatt:  hard, difficult
    helfe:  to help
    immer noch:  still, yet
 es Kaendi:  candy
der Kall (Kaerl):  fellow, chap
der Karrichhof (Kaerrichhof):  churchyard, cemetary
die Kaundi, Kaundis:  county
    koche:  to cook
 es Koches:  cooking
der Kuche:  cake
 es Land, Lenner:  land
    leddich:  single, unmarried
    leige:  to lie (be recumbant)
die Leit:  people
    liewe:  to like, love
die Maetsch:  match
der Manskall, Mannsleit:  man
 es Meedel, Meed:  girl
    meene:  to mean; to think, be of the opinion
    neddemol:  not even
```

oft: often
der Riggelweg: railroad
    runslich: wrinkly, wrinkled
    schaffe: to work
der Schnee: snow
der Schockelschtuhl,-schtiehl: rocking chair
    schockle: to rock
    schpringe: to run, jump; to "go together," date
der Schpringer: baby boy, little felow
die Schtadtleit: city people
    schwer: heavy; difficult, hard
    sex: six
    sehne: to see
die Seit, Seide: side
    siwwe: seven
    sogaar: even
der Tschentelmann, -menner: gentleman
der Urenkel: great grandson
der Urgroossdaadi: great grandfather
    verdien'e: to earn
    verschproch'e: "promised"
die Verwand'e (pl.): relatives
 es Weibsmensch, Weibsleit: woman
    weiss: white
die Witfraa, Witweiwer: widow
     wiescht: ugly
der Zent: cent
    ziehge: to pull, move
    zwelf(e): twelve
der Zwilling, Zwillinge: twins

Vocabulary Notes

backe, denke, draue, helfe, gleiche, koche, schaffe,
    schpringe, sehne, verdiene are all conjugated like
    kumme.

gewwe, glaawe, liewe: conjugated like schreiwe.

schockle: conjugated like rechle.

baalkeppich: There are several variations of this
    word, some of which are baldkeppich, blodkeppich,
    bludkeppich.

es Backes, es Koches: A noun can be made out of a
verb by adding -s to the infinitive. Such a noun
is always neuter, and because it is a noun, it is
always capitalized. Usually it is used with a possessive, meinre Gremmemm ihre Backes "my grandmother's baking."

der Boi: also der Pei.

danke: Also dank(s), dank(i) schee, dank scheene,
grooss Dank.

draue "to trust" and helfe "to help" take dative objects, as explained in the following grammar section.

gaern: This word is used idiomatically with a verb to
mean "like to": Mer gehne gaern ins Schteddel "We
like to go into town"; Lescht du gaern en Buch? "Do
you like to read a book?"

geborni: The family name following geborni is a
woman's leddicher Naame, her "maiden name."

Geschwischder: also Schwischdert.

die Groossmemm: also Groossmammi,-s.

es Land, Lenner: The loss of the d before the plural
ending -er is called assimilation. We say the d is
"assimilated" to the n. It is a relatively common
occurrence in PG; for instance, we learned in Chapter One that the plural of Kind is Kinner.

leige "to lie" is quite irregular in its conjugation;
it will be found in its entirety in the grammar section of this lesson.

Grammar

A. Demonstratives

1. In English, demonstrative adjectives are such
words as "this," "that," "these," "those." Like the E
articles, these adjectives may precede nouns of all
three genders, masculine, feminine, or neuter. But
this is not the case with the PG demonstrative adjective daer (darr) "this." In a sense, these forms are

really accented definite articles; there are, therefore, no unaccented forms.

	Singular			Plural
	M.	F.	N.	All Genders
N. & A.	daer	die	des	die
D.	dem	daerre	dem	denne

2. Just like the possessives, the demonstratives
can be used in place of nouns; they are then called
demonstrative pronouns. And because they are accented
forms, they are often used with inverted word order
for the sake of emphasis:

Ich geb dem Kall ken
Zent; nee, <u>dem</u> gewwich
ken Zent!

I'm not giving that fel-
low a penny; no, <u>him</u> I'm
giving no penny!

Mir gehne net mit denne
Leit; nee, mit <u>denne</u>
gehne mer net!

We're not going with
those people; no, with
<u>them</u> we're not going!

B. The Pronoun <u>ken</u>, <u>kee</u> "none"

1. <u>ken</u>, <u>kee</u> may also be used as a pronoun, in which
instance it has the following forms:

	Singular			Plural
	M.	F.	N.	All Genders
N.& A.	kenner	kenni	kens	kenni
D.	kem	kennre	kem	kenne

<u>kennere</u> is an alternate form of <u>kennre</u>

2. The following sentences illustrate the use of
<u>ken</u> and <u>kenner</u> as adjective and pronoun:

Hoscht du ken Bensil?
Nee, ich hab kenner.

Don't you have a pencil?
No, I have none.

Sie hot ken Buch, un
er hot aa kens.

She has no book, and he
also has none (hasn't
any).

Mir hen ken Kinner, un
sie hen aa kenni.

We have no children, and
they have none either.

C. Verbs

1. If you followed the PG text on the left with the E on the right, then you noticed that several PG verbs in the present tense were translated into the E future tense. Three examples follow:

Ball watt er elder.	Soon he will get older.
Dann gehnich mit denne zwee.	Then I will go with those two.
Dann ziehge sie noch der Schtadt.	Then they will move to the city.

The PG present tense can be used to render the future tense, especially if an adverbial expression of future time is included:

Marriye gehne mer ins Schteddel.	Tomorrow we will go into town.

Those who have a feeling for these things know that "Tomorrow we are going (to go) into town" would also be a good translation. In E, the <u>present progressive</u> can be used instead of the future if we include an adverb of future time.

2. Also, if <u>lang</u> or <u>schunn lang</u> is added to the PG present tense form, then the E present perfect progressive is a good translation:

Sie schpringe schunn lang zamme.	They have been going together for a long time (already).
Er wuhnt schunn lang im Schteddel.	He has been living in the town for a long time.

3. <u>leige</u> "to lie (recline)": The present tense of the verb <u>leige</u> has the following forms:

Singular	Plural
1. ich lei	1. mir,mer leige
2. du leischt	2. dihr,der leit; ihr,er leie; etc.
3. er,sie,es leit	3. sie leige

4. <u>draue</u> "to trust"; <u>helfe</u> "to help": We have seen

that verbs like <u>gewwe</u> "to give" can take both direct and indirect objects:

Ich geb em Mann es Buch.   I give the man a book.

Another group of PG verbs take dative objects (normally the case of the indirect object) instead of the usual accusative objects (normally the case of the direct objects). Two such verbs are used in the preceding reading selection, <u>draue</u> "to trust" and <u>helfe</u> "to help."

Ich drau dem Mann net.   I don't trust that man.

Drauscht du denne Leit?   Do you trust those
                             people?

Mer helfe unserm Paepp.   We're helping our father.

Helft dihr eierer Memm?   Are you helping your mom?

Other such verbs will be pointed out as they appear in subsequent readings.

CHAPTER THREE

Part One

In Chapters One and Two, the second sections were more developed and thus more difficult versions of the first sections. Chapter Three varies this pattern in that its two parts cover two different but related vocabularies. The "meat" of part one is cardinal and ordinal numbers, certainly valuable for everyday use. Our knowledge of numbers will then be used in part two as we study telling time.

Zehe gleene Inschebuwe fische mit der Lein,
Eener fallt ins Wasser, dann waare's yuscht
    noch nein.

Acht gleene Inschebuwe hen die Gens gedriwwe,
Eener beisst der Genserich, dann waare's
    yuscht noch siwwe.

Sex gleene Inschebuwe heere Schpott un Schimf,
Eener schemmt sich halwer dot, dann waare's
    yuscht noch fimf.

Vier gleene Inschebuwe fiedre die Sei,
Eener fallt ins Seifass, dann waare's yuscht
    noch drei.

Zwee gleene Inschebuwe fresse immer mehner,
Eener iss verschprunge, dann waar aa yuscht
    noch eener.[1]

[1]*John Birmelin, from "Ten Little Indians,"* Later
Poems.

55

Nummere   Numbers
Mer Zeehle Enwennich   Let's Count a Bit

A. Cardinal Numbers

1. The numbers we use most often--one, twelve,
forty-five, one hundred, and so on--are called cardinal numbers. The following are the PG cardinal numbers:

		We must take special note
0 (die) Null		of the following differ-
1 eens	11 elf	ences:
2 zwee	12 zwelf	
3 drei	13 dreizeh	
4 vier	14 vazeh	eens elf
5 fimf	15 fuffzeh	zwee zwelf
6 sex	16 sechzeh	vier vazeh
7 siwwe	17 siwwezeh	fimf fuffzeh
8 acht	18 achtzeh	sex sechzeh
9 nein	19 neinzeh	
10 zehe	20 zwansich	

Alternate forms usually used if a number does not
modify a following noun: viere, fimfe, sexe, achte,
neine, elfe, zwelfe.

2. Un nau gehne mer   And now we continue:
   weider:

20 zwansich	60 sechzich	
21 eenunzwansich	66 sexunsechzich	
22 zweeunzwansich	70 siwwezich	
30 dreissich	77 siwweunsiwwezich	
33 dreiundreissich	80 achtzich	
40 vazich	88 achtunachtzich	
44 vierunvazich	90 neinzich	
50 fuffzich	99 neinunneinzich	
55 fimfunfuffzich	100 en hunnert	

Again, note the following differences:

zwee	zwansich
eens	eenunzwansich
vier	vazich
fimf	fuffzich
sex	sechzich

Also, note that <u>zwansich</u> is spelled with one <u>s</u>,
<u>dreissich</u> with <u>ss</u>, and <u>vazich</u>, <u>fuffzich</u>, and so on
with a <u>z</u>.

3. Un nau weider: And now continuing:

en hunnert
en hunnert un eens
en hunnert un zwee (un so weider)

zwee hunnert
drei hunnert (un so weider)

nein hunnert neinunneinzich
en dausend
zwee dausend (un so weider)

nein hunnert neinunneinzich dausend nein hunnert
    neinunneinzich
en millyon'

Un sell iss genunk. Sogaar die Ladderi gebt eem
net meh(ner).

4. If you had 1,979 books in your library, you would
probably say you had one thousand nine hundred and
seventy-nine books. If you bought a new car in 1979,
you would probably say it's a nineteen seventy-nine.
You would, no doubt, use the same expression for the
year 1979.

The PG speaker does exactly the same thing: <u>en</u>
<u>dausend nein hunnert neinunsiwwezich Bicher</u>; <u>es Yaahr</u>
<u>neinzeh hunnert neinunsiwwezich</u>.

B. Ordinal Numbers

1. First, second, third, fourth, and so on, are
called ordinal numbers. To form the ordinals in PG,
one adds the ending -<u>t</u> to the cardinal numbers from 4
through 19 (except 8, which already ends in -<u>t</u>): <u>der</u>,
<u>die,es viert</u>; <u>der,die,es fimft</u>; <u>der,die,es sext</u>; and
so on through <u>der,die,es neinzeht</u>.

2. Starting with 20, one adds the ending -<u>scht</u>:
<u>zwansichscht</u>, <u>eenunzwansichscht</u>, <u>dreissichscht</u>, <u>hun-</u>
<u>nertscht</u>, <u>zwee hunnertscht</u>, <u>vier hunnert vierunvaz-</u>
<u>ischscht</u>, and so on.

3. First, second, third, and eighth are irregular
in PG:

the first	der,die,es aerscht
the second	der,die,es zwett
the third	der,die,es dritt
the eighth	der,die,es acht

Nau kannscht du schee in Deitsch zeehle. Nee? Noch
net? Well, mer kann die Nummere aerscht yuscht lese,
awwer dann muss mer sie auswennich lanne. Noh kann mer
enwennich rechle. Die Andwadde sin rechts driwwe.

Awwer mer soll net gucke, eb mer muss.

1. Du hoscht drei Schweschdere un drei Brieder. Wieviel Geschwischder hoscht du? (Ich hab _____ Geschwischder.)

2. Du hoscht zwee Brieder. Der eent iss vier Yaahr
alt, un der anner iss fuffzeh. Wieviel elder iss er?
(Er iss _____ Yaahr elder.)

3. Du bischt dreizeh Yaahr alt. Dei Bruder iss ee
Yaahr elder. Wie alt iss er? (Er iss _____ Yaahr alt.)

4. Du hoscht zwee Schweschdere. Eeni gebt der (dir)
fuffzeh Zent. Die anner gebt der zwansich Zent. Wieviel Geld hoscht nau? (Nau hawwich _____ Zent.)

5. Du gebscht deinre Schweschder fimfunzwansich Zent
un deim Bruder fuffzeh Zent. Wieviel Geld gebscht ne?
(Ich geb ne _____ Zent.)

6. Du gebscht eem Bruder zwansich Zent un eem dreissich. Wieviel Geld gebscht denne zwee? (Ich geb ne
_____ Zent.)

7. Dei Paepp hot siwwe Geschwischder. Sie sin all
gheiert un hen alle zamme sexunsechzich Kinner. Wieviel Verwande sin des? (Ich zeehl _____ Verwande.)

8. Dei Paepp hot drei Brieder. Eener hot fimf
Kinner, der zwett hot drei, un der dritt hot yuscht
zwee. Wieviel Kinner sin des? (Des sin _____ Kinner.)

Now you can count nicely in German. No? Not yet?
Well, one can first of all just read the numbers, but
then one must learn them by heart. Then one can do a
little math. The answers are over to the right.

But one should not look before one must (has to).

1. You have three sisters and three brothers. How
many siblings (brothers and sisters) do you have? (I
have [sex] siblings.)

2. You have two brothers. The one is four years old
and the other is fifteen. How much older is he? (He
is [elf] years older.)

3. You are thirteen years old. Your brother is one
year older. How old is he? (He is [vazeh] years old.)

4. You have two sisters. One (of them) gives you
fifteen cents. The other gives you twenty cents. How
much money do you have now? (Now I have [fimfundreis-
sich] cents.)

5. You give your sister twenty-five cents and your
brother fifteen cents. How much money do you give
them? (I give them [vazich] cents.)

6. You give one brother twenty cents and one thirty.
How much money do you give these two? (I give them
[fuffzich] cents.)

7. Your dad has seven siblings. They are all married and have all together sixty-six children. How
many relatives does that make? (I count [achtzich]
relatives.)

8. Your dad has three brothers. One has five children, the second has three, and the third has just two.
How many children are they? (That makes [zehe] children.)

9. Du hoscht en Daaler. Du gebscht eenre Schweschder vazich Zent un eenre sechzich. Wieviel Geld hoscht
iwwerich? (Ich hab ____ iwwerich.)

10. Dei Mudder hot drei Kinner, du un dei zwee
Schweschdere. Ze (zu) dritt geht dihr ins Schteddel.
Wieviel Kinner sin noch deheem? (Do sin ____ Kinner
nau deheem.)

Vocabulary

die Andwatt, Andwadde: answer
    auswennich: "by heart"
    auswennich lanne: to memorize
der Daaler: dollar
    dritt: third; zu dritt: the three of us/you/them
    eb: before
    eent: one (of two)
    genunk: enough
    iwwerich: left (over)
    mer: one, they, people (see note)
    misse: must, to have to
    nau: now
    noch: still, yet; noch net: not yet
    noh: then
die Nummer, Nummere: number
    schee (scheen+ending): nice(ly)
    solle: should
    weider: on, further, farther
    wieviel?: how many, how much?
 es Yaahr, Yaahre: year
    zeehle: to count

Vocabulary Notes

mer: The indefinite pronoun mer, meaning "one," "they,"
    "people," has only a nominative singular form. It
    "borrows" its dative form eem from the pronoun eener
    "one," whose declension will be found in the follow-
    ing grammar section.

gucke and zeehle: conjugated like kumme.

Daaler, Yaahr, Zent, Daag: After numerals, these words
    generally appear in the singular. Other such words
    will be pointed out as they appear in subsequent vo-
    cabularies.

9. You have a dollar. You give one sister forty cents and one sixty. How much money do you have left? (I have [nix] or [ken Geld] left over.)

10. Your mother has three children, you and your two sisters. The three of you go into town. How many children are still at home? (There are [ken, kee] children at home.)

Grammar

A. The Adjective and Pronoun ee(ns)

1. When used as an adjective, the cardinal number eens has the following forms:

	M.	F.	N.
N. & A.	ee	ee	ee
D.	eem	eenre	eem

2. Like the possessives and demonstratives, this numeral can be used also as a pronoun:

	M.	F.	N.
N. & A.	eener	eeni	eens
D.	eem	eenre	eem

The following sentences illustrate the use of both the adjective and the pronoun:

Hoscht Gschwischder?
Ya, ich hab ee Bruder
un ee Schweschder.

Do you have siblings?
Yes, I have one brother
and one sister.

Hoscht en Bruder?
Ya, ich hab eener.

Do you have a brother?
Yes, I have one (m.).

Hoscht en Schweschder?
Gewiss, ich hab eeni.

Do you have a sister?
Yes, I have one (f.).

Sehnscht en Buch uff em Desk? Ya, ich sehn eens.

Do you see a book on the desk? Yes, I see one (n.).

B. The indefinite pronoun mer "one"

1. A vocabulary note tells us that the indefinite pronoun mer "one" has but a nominative form; eem, the

dative of eener, takes the place of the missing dative
of mer:

Mer weess, viel Leit	One knows, many people
helfe eem gaern.	help one gladly.

"They" and "people" are also used in English in such
a way that mer would be a correct equivalent (translation):

In daerre Klass do	In this (here) class,
schwetzt mer yuscht	people (or they) speak
Deitsch.	only German.

2. Thus the PG word mer can mean several things, but
its use within the context of a sentence will make
clear exactly what it means.

Mer kumme gewehnlich	We usually come at three
um drei Uhr.	o'clock.

Mer must be the first person plural pronoun "we,"
because it is followed by the first person plural verb
form kumme.

Mer kummt gewehnlich	One usually comes at
um drei Uhr.	three o'clock.

Mer must be the indefinite pronoun "one," because
it is followed by the third person singular kummt.

Er gebt mer en Daaler.	He gives me a dollar.

The mer in this sentence must be the dative of ich;
that is, it must be the indirect object following the
verb gewwe "to give."

3. One other point: mer meaning "we" or "to me" has
the accented form mir, as in:

Mir gehne ins Schteddel.	We are going into town.
Er gebt mir en Daaler.	He gives me a dollar.

The indefinite pronoun mer has no such accented
form; it is the PG dialect equivalent of the Standard
German indefinite pronoun man.

CHAPTER THREE
Part Two

Count yourself lucky if there is one of those play
clocks around the house, the ones we buy children to
teach them how to tell time. And for the next week or
so, be a clock-watcher--in PG, of course.

Die alt Uhr henkt datt an der Wand,
Ihr Gsicht iss mer gans gut bekannt,
Sie gnackt noch wie in frieh'rer Zeit,
Un saagt zum Mensch, halt dich bereit,
Gnick, gnack,
Gnick, gnack,
Vun Schtunn zu Schtunn eilt hie die Zeit
Un draagt uns noch der Ewichkeit.[1]

[1]*Isaac S. Stahr, from "Die Alt Uhr,"* PG Verse.

63

Was Zeit iss es?

Weescht du was mer saagt, wann ebber froogt, "Was Zeit iss es?" ("Wieviel Uhr iss es?") Nee, du kannscht em net saage? Dann kenne mer heit ebbes lanne.

Sehnscht die Uhr do? Die hot en Uhregsicht, odder en Zifferblaat, mit Ziffere odder Nummere druff--eens bis zwelfe. Mer sehnt aa die zwee Zoiyer, odder Hend. Deel Leit saage Zeeche, die Uhrezeeche (sing. es Zeeche).

Der eent Zoiyer (die eent Hand) iss lang, odder grooss. Er (sie) deidt uff die Minudde, un mer heesst en (sie) der Minuddezoiyer. Der anner Zoiyer iss kaz, odder glee, un daer deidt uff die Schtunne. Mer heesst ihn der Schtunnezoiyer.

Aerscht drehe mer alle zwee (alle beed) uff der Zwelfder. Nau iss es zwelf Uhr. Um zwelf Uhr kann es Middaag sei odder Halbnacht sei. (Mer saagt aa Midnacht odder Middernacht.) Iss es hell draus? Dann iss es Middaag. Iss es dunkel draus? Dann iss es nadierlich Halbnacht.

In eenre Minutt iss es dann ee Minutt nooch de zwelfe, odder ee Minutt nooch zwelf Uhr (mer saagt aa iwwer zwelfe), un in noch eenre iss es zwee Minudde nooch de zwelfe, odder nooch zwelf Uhr.

Mer schtelle der Minuddezoiyer uff der Eender, un es iss fimf Minudde nooch de zwelfe; uff der Zwedder, zehe Minudde nooch de zwelfe; uff der Dridder, fuffzeh Minudde nooch de zwelfe. Awwer gewehnlich saagt mer, es iss (en) Vaddel nooch zwelfe (iwwer zwelfe). Fuffzeh Minudde sin en Vaddel Schtunn.

Nau schtelle mer der glee Zoiyer, odder Schtunnezoiyer, uff der Vierder, un der grooss Zoiyer uff der Sexder. Mer kann dreissich Minudde nooch viere saage, awwer des iss die Helft vun sechzich Minudde, die Helft vun eenre Schtunn. Un des iss halbwegs vun viere bis fimfe. So saagt mer gewehnlich, es iss halwer fimfe. Mer schtelle der glee Zoiyer uff der Achtder. Nau iss es halwer neine; uff der Neinder, halwer zehe; uff der Zeheder, halwer elfe; un so weider.

What Time is it?

Do you know what one says when someone asks, "What time is it?" (lit. "How much o'clock is it?") No? You can't tell him? Then we can learn something today.

Do you see this clock here? It has a clock face, or dial, with numerals or numbers on it--one to twelve. One also sees two pointers, or hands. Some people say Zeeche, the clock Zeeche (synonymous with Zoiyer).

The one pointer (the one hand) is long, or big. It (it f.) points to the minutes, and one calls it the minute hand. The other hand is short, or small, and this (one) points to the hours. One calls it the hour hand.

First (of all) we turn the two to the twelve. Now it is twelve o'clock. At twelve it can be midday or midnight. (Midnacht and Middernacht are synonymous.) Is it light outside? Then it is midday. Is it dark outside? Then it is naturally midnight.

In one minute it will be one minute after twelve, or one minute after twelve o'clock (one also says past twelve), and in one more it will be two minutes after twelve, or after twelve o'clock.

We place the minute hand on the one and it is five minutes after twelve; on the two, ten minutes after twelve; on the three, fifteen minutes after twelve. But usually one says it is (a) quarter after twelve (past twelve). Fifteen minutes are a quarter hour.

Now we place the little hand, or hour hand, on the four, and the big hand on the six. One can say thirty minutes after four, but that is the half of sixty minutes, the half of one hour. And that is halfway from four to five. So one usually says it is half (of the way to) five. We place the little hand on the eight. Now it is half nine (half past eight); on the nine, half past nine; on the ten, half past ten; and so on.

Mer schtelle der grooss Zoiyer uff der Siwweder: es iss fimfunzwansich Minudde bis zwelfe, odder ver zwelfe, odder ver de zwelfe. Der grooss Zoiyer uff der Neinder: es iss nau fuffzeh Minudde bis zwelfe (un so weider). Mer kann awwer aa des saage: Es iss (en) Vaddel bis zwelfe (un so weider). Zehe Minudde schpeeder iss es fimf Minudde ver zwelfe, un uff ee mol iss es widder zwelf Uhr.

Kannscht nau die do zehe Frooge andwadde?

1. Wieviel Segunde hot en Minutt? Hen fimf Minudde? (Ee Minutt hot _____ Segunde; fimf Minudde hen _____ Segunde.)

2. Wieviel Minudde sin in re Schtunn? (In eenre Schtunn sin _____ Minudde.)

3. Wieviel Schtunne sin imme Daag? (In eem Daag sin _____ Schtunne.)

4. Wie lang schafft mer gewehnlich in eem Daag? (In eem Daag schafft mer _____ Schtunne.)

5. Wie lang schafft mer in eenre Woch? (In eenre Woch schafft mer _____ Schtunne.)

6. Um sex Uhr gehscht in die Schul. Um siwwe Uhr bischt datt. Wie lang nemmt des? (Des nemmt _____ _____.)

7. Um acht Uhr gehscht ins Schteddel. Des nemmt fuffzeh Minudde. Was Zeit iss es nau? (Es iss fuffzeh Minudde _____ _____ _____; es iss nau (____) _____ _____ (____) _____.)

8. Um drei Uhr gehscht zu de Groosseldre uff Bsuch. Des nemmt dreissich Minudde. Wieviel Uhr iss es nau? (Es iss _____ _____ _____ _____; es iss _____ _____.)

9. Der Minuddezoiyer iss graad uff em Achtder, un der Schtunnezoiyer iss beinaah uff em Neinder. Was Zeit iss es? (Es iss _____ _____ _____ (or ____) _____.

10. Wann gehscht marriyeds schaffe odder in die Schul? Wann bischt oweds odder nammidaags widder deheem?

We place the big hand on the seven: it is twenty-
five minutes to twelve, or before twelve, or before the
twelve. The big hand on the nine: it is now fifteen
minutes to twelve (and so forth). But one can also say
this: It is (a) quarter to twelve (and so on). Ten
minutes later it will be five minutes to twelve, and
all of a sudden it is again twelve o'clock.

Can you now answer these (here) ten questions?

1. How many seconds does a minute have? Do five minutes have? (One minute has [sechzich] seconds; five
minutes have [drei hunnert] seconds.)

2. How many minutes are in an hour? (In one hour
are [sechzich] minutes.)

3. How many hours are in a day? (In one day are
[vierunzwansich] hours.)

4. How long does one usually work in one day? (In
one day one works [acht] hours.)

5. How long does one work in one week? (In one week
one works [vazich] hours.)

6. At six o'clock you go to school. At seven o'clock
you are there. How long does that take? (That takes
[ee Schtunn] or [sechzich Minudde].)

7. At eight o'clock you go into town. That takes 15
minutes. What time is is now? (It is fifteen minutes
[nooch acht Uhr]; it is now [(en) Vaddel nooch (de)
achte].)

8. At three o'clock you go to visit the grandparents.
That takes thirty minutes. What time is it now? (It
is [dreissich Minudde nooch drei]; it is [halwer viere].)

9. The minute hand is right on the eight, and the
hour hand almost on the nine. What time is it? (It is
[zwansich Minudde ver (or bis) neine].)

10. When do you go to work or school in the mornings? When are you home again in the evenings or afternoons?

Vocabulary

    all(e): all
    alle beed: both
    alle zwee: both (all two)
    beinaah: almost, nearly
    bis: to, till; as far as
    deide: to point
    des: that
    do: here
    draus: outside
    drehe: to turn
    dunkel: dark
    es: it
    gewehn'lich: usually
    graad: exactly, right; straight
die Halbnacht: midnight
    halbwegs: half way
    halwer: half
    heit: today
die Helft, Helfde: half
    hell: light, bright
    iwwer: over, past; iwwer zwelfe: past twelve
 es Middaag, -daage: midday
die Middernacht: midnight
die Midnacht: midnight
der Minud'dezoiyer: minute hand
die Minutt', Minudd'e: minute
die Nacht, Nachde (Nechde): night
    noch: still, yet
    noch eener: one more
    nooch: after
    oweds: in the evening, evenings
    schpeeder: later
    schtelle: to place, put, set; die Uhr schtelle:
        to set the clock, watch
die Schtunn, Schtunne: hour
der Schtunnezoiyer: hour hand
die Segund': second
    uff eemol: all of a sudden
die Uhr, Uhre: clock, watch
    wieviel Uhr: what time
die Uhregsicht: face of clock, watch
    um: around; um vier Uhr: at four o'clock
 es Vaddel: quarter

weider: on, further, farther; un so weider: and
    so forth
der Ziffer, Ziffere: number, figure
 es Zifferblaat: dial (face) of clock, watch
der Zoiyer: hand of clock, watch

Vocabulary Notes

drehe, schaffe, and schtelle: conjugated like kumme.

des: can refer to any gender, any number, if it precedes a linking verb: Des iss mei Vadder, des iss
mei Mudder, des sin mei Eldre.

um means "around," unless it is used with time, in
which case it means "at": um drei Uhr "at three
o'clock."

Grammar

A. Number Names

1. With few exceptions, the names of numbers (always masculine) derive from the cardinal number and
the ending -der: der Vierder, Fimfder, Sexder, Siwweder, etc.

2. The exceptions:

eens	der Eender
zwee	der Zwedder
drei	der Dridder

B. Fractions

1. With few exceptions, fractions to twentieth are
formed by adding -del to the cardinal numbers: es
Fimfdel, es Sexdel, es Acht(d)el, es Zehedel, etc.

2. The exceptions: halb (halw+ending) "half," die
Helft, Helfde "the half," es Driddel "third," es Vaddel "fourth."

3. Starting with twentieth, -schdel is added to the
cardinal numbers: es Zwansichschdel, es Dreissich-
schdel, es Vazichschdel, etc.

CHAPTER FOUR

Part One

Before starting this reading selection, it would be
wise to study the names of the seven days of the week:
Sunndaag, Muundaag (Moondaag), Dinschdaag (Diensch-
daag), Mittwoch, Dunnerschdaag, Freidaag, Samschdaag.
We will study the various parts of the day in the
second part of this chapter.

Wesch am Mundaag!
Dinschdaag kumme!
Gschickt schee Neehes
Mittwoch kumme.
Dunnerschdaags dann
Geht's ans Butze,
Freidaags aa noch
Dreck verdreiwe.
Back am Samschdaag!
Noh bringt Sunndaag
Schul un Karrich. [1]

[1] *Russell W. Gilbert, from "Re Hausfraa Ihre Arrewet,"*
Bilder un Gedanke.

71

Die Woch

Die Woch hot siwwe Daag. Der aerscht Daag vun der
Woch iss Sunndaag un an sellem Daag gehne viel Leit in
die Karrich. Deel gehne marriyeds, deel oweds. Sie
kenne dann ihre scheene Sunndaagsgleeder aaduh, un
sell gleiche sie. Sunndaags gehne die Kinner aa in
die Sunndaagsschul, un do misse sie aa gude Gleeder
aaduh. Des gleiche sie net immer, abbaddich die Buwe.

Nammidaags sitzt mer rum, geht uff Bsuch, odder
grickt Bsuch; es iss yo en Ruh(k)daag. Awwer Sunndaag
oweds soll mer net zu schpeet, awwer schee frieh, ins
Bett geh. Am neegschde Marriye muss mer widder frieh
uffsei, un sei Schaffgleeder odder Waerdaagsgleeder
aaduh. Waerdaags muss mer schaffe geh.

Un deswegge meene viel Leit, der Muundaag iss der
aerscht Daag vun der Woch. Ya, der Muundaag iss en
Schaffdaag fer viel Leit, un gude, ehrliche Schaffleit
misse widder hatt schaffe. Fer viel Weibsleit iss es
aa der Weschdaag; der alt Weschkarreb iss widder voll
mit Wesch. Un fer die Kinner iss der Muundaag aa en
Schuldaag; selli misse zrick in die Schul.

Der neegscht Daag iss Dinschdaag, der dritt Daag in
der Woch. Dann kummt nadierlich der Mittwoch. Nau
simmer mittwegs darrich en gansi Woch. Nooch em Mittwoch kummt der fimft Daag, der Dunnerschdaag; un dann,
Gott sei dank, iss es Freidaag.

Bezaahlsdaag! Nau griege mer widder Geld. Un viel
Leit kenne Freidaag oweds schpot ins Bett. Sie brauche
Samschdaag marriyeds net so frieh uffschteh. Sie kenne
Freinde bsuche, danse geh, odder in re Barschtubb rum-
hocke. Des geht Schtunnelang so weider bis lang nooch
Halbnacht, un wann die Leit heemkumme, do funkle schunn
die Schtanne am schwaze Himmel. Awwer mer brauch so
ebbes, abbaddich nooch re ganse Woch.

Noh iss es Samschdaag. Uff re Bauerei gebt es immer
ebbes zu duh, graad wie amme Waerdaag. Awwer uff
sellem Daag bleiwe etliche Leit deheem, odder faahre
ins Schteddel odder sogaar in die Schtadt. Datt kaafe
sie ihre Esssache un verleicht neie Gleeder. Algebott
gleicht mer ebbes Neies un Scheenes. Es muss nix
Dei(e)res sei.

The Week

The week has seven days. The first day of the week
is Sunday and on that day many people go to church.
Some go in the morning, some in the evening. They can
then put on their nice Sunday clothes, and they like
that. The children go to Sunday school, and they also
have to put on good clothes. They don't always like
that, especially the boys.

In the afternoon one sits around, goes visiting or
gets company (visitors); it is, of course, a day of
rest. But Sunday evenings one should go to bed not
too late, but nice and early. On the next morning one
must again be up early and put on his work clothes or
weekday clothes. On weekdays one has to go to work.

And for that reason many people think Monday is the
first day of the week. Yes, Monday is a work day for
many people, and good, honest working people must work
hard again. For many women it is also the washday; the
old wash basket is again full of wash. And for the
children Monday is also a school day; they must (go)
back to school.

The next day is Tuesday, the third day of the week.
Then comes naturally Wednesday. Now we are midway
through a whole week. After Wednesday comes the fifth
day, Thursday; and then, thank God, it is Friday.

Payday! Now we get money again. And many people
can go to bed late on Friday nights. They don't have
to get up early on Saturday mornings. They can visit
friends, go dancing, or sit around in a bar room. That
goes on (continues) for hours until long after midnight,
and when the people come home, the stars are already
sparkling in the black heavens. But one needs something like that, especially after a whole week.

Then it is Saturday. On a farm there is always
something to do, just like on a weekday. But on that
day some people stay at home, or drive into town or
even into the city. There they buy their food and maybe new clothes. Every once in a while one likes something new and nice. It doesn't have to be anything
expensive.

Un noch en paar gude Wadde:

Mer saage mol, heit iss Mittwoch. Geschder waar es Dinschdaag, un vorgeschder waar es Muundaag.

Marriye iss es Dunnerschdaag, un iwwermarriye iss es dann Freidaag.

En Woch zrick (heit acht Daag zrick) waar es Mittwoch, un neegscht (neegschdi) Woch (heit iwwer acht Daag) iss es widder Mittwoch.

Un en Woch dernoh, heit iwwer vazeh Daag, iss es aa widder Mittwoch.

Vocabulary

    aaduh: to put on, get dressed
die Barschtubb, -schtubbe: barroom
 es Bett, Bedder: bed; ins Bett geh: to go to bed
    bezaahl'e: to pay
der Bezaahls'daag, -daage: payday
    brauche: to need, have need of
der Bsuch: visit, "company"; uff Bsuch geh: to go
      visiting; Bsuch griege: to get company
    bsuche: to visit
    danse: to dance
    deier: expensive; nix Deieres: nothing expensive
    deswegge: therefore, for that reason
    faahre: to drive, to ride in a vehicle
der Freind, Freinde: friend
    funkle: to sparkle
    geh: to go; danse geh: to go dancing; schaffe
      geh: to go to work
    geschder: yesterday
    gewwe: to give; es gebt: there is, there are
die Gleeder: clothes
der Gott, Gedder: God; Gott sei dank: thank God
    griege: to get, receive
    heemkumme: to come home
der Himmel: sky, heaven
    iwwermarriye: day after tomorrow
die Karrich, Karriche (Kaerrich, Kaerriche): church
der Marriye: morning
    mittwegs: midway

And a few more good words:

Let's say today is Wednesday. Yesterday it was
Tuesday, and the day before yesterday it was Monday.

Tomorrow it is (going to be) Thursday, and the day
after tomorrow it will then be Friday.

A week ago (today eight days ago) it was Wednesday,
and next week (today over eight days) it will again be
Wednesday.

And a week after that, today in fourteen days, it
will also be Wednesday again.

der Nammidaag: afternoon
    nammidaags: afternoons, in the afternoon
    nei: new; ebbes Neies: something new
der Ruh(k)daag, -daage: day of rest
    rumhocke: to sit around
der Schaffdaag, -daage: work day
die Schaffgleeder: working clothes
die Schaffleit: working people
    schee (scheen+ending): nice, beautiful
        ebbes Scheenes: something nice
    schpot: late
der Schtann, Schtanne: star
    schtunnelang: for hours
    schwaz: black
    sunndaags: on Sundays
die Sunndaagsgleeder: Sunday clothes
die Sunndaagsschul: Sunday school
    uffschteh: to get up
    vorgeschder: day before yesterday
der Waerdaag, -daage: weekday
die Waerdaagsgleeder: weekday clothes
der Weschdaag, -daage: washday
    wesche: to wash
der Weschkarreb: wash basket
die Woch, Woche: week
    yo: affirmative answer (yes) to a negative ques-
        tion, but also a particle (see note below)

Vocabulary Notes

bsuche, faahre, mache, meene, schaffe: conjugated
like kumme.

danse, wesche: conjugated like schwetze.

funkle: conjugated like rechle.

griege: This verb is not conjugated like saage; it
will be taken up in the grammar section of this
lesson.

misse and solle are modal auxiliaries; they too will
be taken up in the grammar section of this lesson.

deswegge: also desweege.

gewwe, es gebt: In the E expressions "there is,"
"there are," (There is a man in the room, there are
two men in the room), the word "there" is not an adverb, as in the expression "over there," but an expletive. The PG equivalent is the expletive es and
the verb gewwe. Because es is always third person,
singular, the verb must always be singular whether
or not a singular or plural form follows: es gebt
siwwe Daag in eenre Woch "there are seven days in
one week."

schee: If schee precedes an adjective or adverb, the
E equivalent is usually "nice and": schee grooss
"nice and big," schee warrem "nice and warm."

yo: We know that yo is used in place of ya for an
affirmative answer to a negative question. But yo
is also a particle like denn; it sometimes defies
translation. It could be used to indicate surprise;
Do sin sie yo! "Well, here they are!" (The implication is that the speaker is surprised to see them.)
It is also used to emphasize a fact: Heit misse mer
schaffe; es iss yo en Schaffdaag. We have not yet
studied commands, but when we do, we will be able to
use this particle again to emphasize a negative command: Duh des yo net widder! "Don't you dare do
that again!"

der Nammidaag: also Nummidaag.

nammidaags: also nummidaags.

Grammar

A. Adjectives

1. In English, adjectives are generally used after linking verbs (e.g., be, become) or immediately before nouns. Adjectives that follow linking verbs are called predicate adjectives; they modify (tell us something about) the subject of the linking verb:

The school is large.   The schoolroom is small.

In the first sentence, "large" follows the verb but describes the school: it is a large school.   In the second sentence, the predicate adjective "small" tells us that the room is a small room.

2. Adjectives placed immediately in front of nouns are called attributive adjectives: a large school, a small schoolroom, an old schoolhouse.   "Large," "small," and "old" are attributive adjectives.

3. PG adjectives are used in exactly the same way.

Die Schul iss grooss.   Die Schulschtubb iss glee.

Grooss and glee are predicate adjectives following the linking verb sei.   In PG, predicate adjectives never take endings!

4. However: en groossi Schul, en gleeni Schulschtubb, en alt (aldes) Schulhaus.   Groossi, gleeni, and alt (aldes) are attributive adjectives.   Note that they have what are called adjectival endings.   Let us now take a close look at them.

5. In this section of Chapter 4, we will consider two adjective declensions, or sets of adjectival endings: (1) the weak declension (weak endings) used with adjectives that are themselves preceded by a definite article (der, die, es), or by the demonstrative adjective daer (die, des), and (2) the mixed declension (mixed endings) used with adjectives that are preceded by an indefinite article (en) or a possessive adjective (mei, dei, etc.).

B. The Weak Declension

Singular

	M.	F.	N.
N.& A.	der alt Mann	die yung Fraa	es glee Kind
D.	em alde Mann	der yunge Fraa	em gleene Kind

Plural
All Genders

N.& A.   die alde (yunge,gleene) Menner (Weiwer,Kinner)
D.   de alde (yunge,gleene) Menner (Weiwer,Kinner)

1. We could spend our time studying each individual ending--no doubt resulting eventually in our memorizing the endings. But let's take a look at "the big picture" and draw three general conclusions:

(1) All the singular nominative and accusative endings, no matter what the gender, have no endings;

(2) All the dative singular forms have an -e ending;

(3) All plurals have an -e ending.

In fact, you may wish to boil it down even further: all singular nominative and accusative have no endings; all others have an -e ending. It's as simple as that!

C. The Mixed Declension

Singular

	M.	F.	N.
N.& A.	en alder Mann	en yungi Fraa	en glee(nes) Kind
D.	me alde Mann	re yunge Fraa	me gleene Kind

Plural
All Genders

N.& A.   ken alde (yunge,gleene) Menner (Weiwer,Kinner)
D.   ken alde (yunge,gleene) Menner (Weiwer,Kinner)

1. Once again, we can see that the dative singular and all the plurals have an -e ending. But the singular N. and A. shouldn't give us much trouble either: -er reflects masculine der; -i reflects feminine die; and -es reflects neuter es, although the adjective without an ending is much more common.

2. Here are a few important points to remember when adding adjectival endings.

(a) Final <u>b</u> becomes <u>w</u>: <u>Der Hund iss lieb; er iss en liewer Hund.</u>

(b) Final <u>t</u> becomes <u>d</u>: <u>gut</u>, <u>gude</u>.

(c) <u>schee</u> and <u>glee</u> (others will be pointed out as they appear in the text) add -<u>n</u> before endings: <u>die schee Fraa, en scheeni Fraa; der glee Bu, en gleener Bu.</u>

(d) Two or more attributive adjectives modifying the same noun have the same endings: <u>gude, ehrliche Schaffleit.</u>

3. PG adjectives may be used as nouns. They must then be capitalized and must retain their appropriate endings: <u>der Alt</u>, <u>en Alder</u>; "the old man, an old man." But when these adjectives-turned-nouns follow <u>ebbes</u> or <u>nix</u>, they are always given the ending -<u>es</u>: <u>ebbes Neies, nix Deieres</u>; "something new, nothing expensive."

PG speakers do not always use adjectival endings on <u>neegscht</u> "next" and <u>letscht</u> "last" when they are followed by time expressions: <u>neegscht Woch</u> (<u>neegschdi Woch</u>), <u>letscht Woch</u> (<u>letschdi Woch</u>).

If an adjective ends in -<u>er</u> or -<u>el</u>, the <u>e</u> is often dropped when adjectival endings are added: <u>des Haus iss deier, es iss en deires Haus; die Nacht iss dunkel, es iss en dunkli Nacht.</u>

D. Adverbs

Adverbs modify verbs, adjectives, and other adverbs. In PG, adverbs do not take endings:

> Sie schafft hatt.   She works hard.

<u>Hatt</u> is an adverb modifying the verb <u>schafft</u>, telling us <u>how</u> she worked.

> Mer soll schee frieh ins   One should go to bed nice
> Bett geh.   and early (nicely early).

E. Verbs

1. Present Tense of Modal Auxiliaries

By this time we should be familiar with the modal auxiliary <u>kenne</u>. We know that its singular forms do

not follow the usual rules as, for instance, kumme
does: the stem vowel e is changed to a, and the first
person and third person forms have no endings.

Also, we know that the plural forms are quite
"regular"; that is, the stem vowel e is retained and we
add the usual endings for all three persons.

There are five more modal auxiliaries just like
kenne, but we will study only two of them in this section of Chapter 4: misse "to have to, must"; and solle
"to be supposed to, should." In addition, we will take
up brauche "to need," a verb that traditionally has not
been called a modal auxiliary, but is used exactly as
the modals are used, either by themselves or with dependent infinitives.

Because of changes in stem vowels, the entire conjugation of each verb follows:

(a) misse "to have to, must"

Singular	Plural
1. ich muss	1. mir,mer misse
2. du musscht	2. dihr,der misst; ihr,er misse, etc.
3. er,sie,es muss	3. sie misse

(b) solle (selle) "to be supposed to, should"

Singular	Plural
1. ich soll	1. mir,mer solle (selle)
2. du sollscht	2. dihr,der sott (sett); ihr,er solle (selle); etc.
3. er,sie,es soll	3. sie solle (selle)

If you rely on your knowledge of how kenne is used,
then you won't have any trouble with misse and solle
(selle).

Er kann Deitsch schwetze.	He can speak German.
Er muss Deitsch schwetze.	He must speak German.
Er soll Deitsch schwetze.	He should speak German; he is supposed to speak German.

Once again, it is important to remember that anything we want to say in addition to the auxiliary verb
and the infinitive must be "sandwiched" between the
two:

Er kann, muss, soll Muun-daag oweds vun siwwe bis halb zehe Deitsch schwetze.	He can, must, should talk German on Monday evenings from seven to nine-thirty.

An answer to a question containing both a modal and a dependent infinitive may drop the dependent infinitive:

Musscht du des Buch lese?	Must you (do you have to) read this book?
Ya, ich muss.	Yes, I must.

As a matter of fact, PG speakers rarely use a dependent infinitive if it is obvious that "go" (or any other verb implying "going") is meant:

Musscht (du) ins Schted-del?	Must you (go) into town?
Ya, ich muss.	Yes, I must (go).

(c) <u>brauche</u> (<u>breiche</u>) "to need, have need of"

Singular	Plural
1. ich brauch	1. mir,mer brauche (breiche)
2. du brauchscht	2. dihr,der braucht (breicht); ihr,er brauche (breiche); etc.
3. er,sie,es brauch	3. sie brauche (breiche)

Like the modals, <u>brauche</u> (<u>breiche</u>) may be used by itself, without a dependent infinitive, in which case it means to need:

Der Schuler braucht en Bleibensil un Babier.	The pupil needs a pencil and paper.

But the verb <u>brauche</u> may be used to mean "not to be required to" if it is accompanied by <u>net</u> and a dependent infinitive:

Die gleene Schieler brauche (breiche) net mit Fedder un Dinde schreiwe.	The little pupils are not required to write (do not have to write) with pen and ink.

2. <u>griege</u> "to get, obtain"

<u>griege</u> would appear to be conjugated like <u>leige</u>, but there are significant differences in the singular forms:

Singular	Plural
1. ich grick	1. mir,mer griegge
2. du grickscht	2. ihr,er griegge; dihr,der grickt, etc.
3. er,sie,es grickt	3. sie griegge

3. The Past Tense of sei "to be"

The verb sei is the only PG verb that has a simple past tense (E "was, were").

Singular	Plural
1. ich waar	1. mir,mer waare
2. du waarscht	2. ihr,er waare; dir,der waart, etc.
3. er,sie,es waar	3. sie waare

The past tense of sei is used exactly as E speakers use the past tense of "to be":

Ich waar im Schteddel.	I was in town.
Waart dihr (waare ihr) ans Kunkels?	Were you at Kunkel's (house, place)?

The past progressive is formed just like the present progressive, but with the past tense forms of sei:

Mei Fraa waar am Wesche.	My wife was doing the wash.
Sie waare all am Danse.	They were all dancing.

4. Double Infinitives

geh is one of a small number of verbs that can take dependent infinitives very much like modal auxiliaries. Perhaps we will understand this construction better if we use our knowledge of modals and their dependent infinitives, e.g., danse kenne "to be able to dance"; schaffe misse "to have to work":

Ich kann denowed danse.	I can dance this evening.
Ich geh denowed danse.	I (am) go(ing) dancing this evening.
Ich muss marriye schaffe.	I must work tomorrow.
Ich geh marriye schaffe.	I'm going to work tomorrow.

If we use a double infinitive such as danse geh or schaffe geh with a modal auxiliary, then we simply treat the entire double infinitive as if it were just one unit, a single infinitive:

Ich kann denowed danse geh.

I can go dancing this evening.

Ich muss marriye schaffe geh.

I must go to work tomorrow.

CHAPTER FOUR

Part Two

Are we ready to think in PG? This next reading selection contains vocabulary of the various parts of the
day, but it may also set us thinking. Why does the
woman always look out the window--even when having supper with her family? She says she doesn't like to gossip on the phone. Would the women who call her up
agree? Does she really work as hard and long as she
claims? Was this truly a glorious day?

"O, geh!"
"So waahr ass ich do schteh!"
"Un waer hoscht gsaat, ass hett's der gsaat?"
"Des weess ich selwer nau net graad.
Doch waardt emol--nau loss mich denke--
O ya! Ich waar do yetz ans Schwenke;
Die Schwenksen hot's woll vun der Moyern
Un die Butzen vun der Boyern
Un die Boyern vun der Kutzen--"
"Un hot die Moyern des gewisst
Un mir nix gsaat?--Mer meent, mer misst--"
"O, well! So iss es mit dem Gschwetz,
's iss immer aerryeds ebbes letz--
"Doch bischt du nau die aerscht die's weess--
Un gell, du bischt mer nau net bees?"
"Ya, well, gut bei!
's iss arrig mit der Retscherei!"
"Ich middel mich mol gaar net nei.--
Vergess mich net, wann d' ebbes heerscht!"
"Verloss dich druff, du bischt die aerscht!" [1]

[1]*John Birmelin, from "Die Retscherei,"* Gezwitscher.

En Hallicher Daag

Es iss gans frieh marriyeds, noch gans finschder, awwer net lang eb es hell watt, eb Daageshelling. En Weil zrick waar der Millichmann do; die Millichboddle rabble immer, un sell macht mich dann wacker. Seller Kall kummt net alle Marriye, yuscht dreimol die Woch, awwer des iss drei mol zu oft.

Der Moond (Muund) waar graad am Unnergeh, un nau iss die Sunn ball am Uffgeh, awwer es iss noch net gans hell. Un do hock ich am Breckfeschtdisch un hab mei Marriyedsesse, guck aus em Fenschder un sehn der Marriyeschtann gans glitzerich drowwe im Himmel. En Nochber geht verbei mit seim Hund. Seller Hund iss awwer lieb.

Die aerschde Schtraahle vun der Sunn kumme graad iwwer der Barrick un ich bin schunn am Schaffe--butze, wesche, un koche. Ich guck als naus aus em Fenschder. Was mache denn die Nochbere datt driwwe? Un waer waar seller Alt datt draus?

So an die zehe Uhr ess ich ebbes, un drink en Koppche Kaffi. Marriyeds gleich ich mei Kaffi, un demarriye drink ich sogaar zwee Koppcher. Un dann iss die Foohn am Glingele. Des iss widder die Moyern, die Retsch. Die weess alles, un sie ruft mich uff un saagt mer alles. Ich gleich sell gaar net, so bleib ich yuscht en Schtunn uff der Foohn.

Der gans Marriye schaff ich hatt, un eblang iss en halwer Daag verbei.

Die Middaagsschtunn iss schunn do, un ball iss es Middaag(esse) faddich. Noh hock ich widder do un ess mei Dinner. Ich guck aus em Fenschder, un der Briefdreeger (Meelmann) iss am Verbeigeh. Wann esst seller Kall eenihau? Un er waar schunn zwee Daag net bei mer. Ferwas dutt ennichebber mir net schreiwe? Well, ich brauch aa niemand net schreiwe.

Glei nooch Middaag muss ich widder schaffe; do iss immer ebbes zu duh. Die Foohn ringt noch ee odder zwee mol. Ferwas saage sie mir immer alles? Sie wisse, ich babbel net gaern am Foohn.

Un so geht's der gans Nammidaag weider. Nau geht's uff drei Uhr, un ich bin als noch am Schaffe. Awwer

A Glorious Day

It is very early in the morning, still very dark,
but not long before it gets light, before dawn. A
while ago the milkman was here; the milk bottles always rattle, and that wakes me up. That fellow
doesn't come every morning, just three times per week,
but that is three times too often.

The moon was just going down, and now the sun is
soon coming up, but it is not yet entirely light. And
here I sit at the breakfast table and have my breakfast, look out the window and see the morning star all
glittery up there in the sky. A neighbor goes by with
his dog. My, but that dog is dear.

The first rays of the sun are just coming over the
mountain and I am already working--cleaning, washing,
and cooking. I always keep looking out the window.
What are those neighbors doing over there? And who was
that old man out there?

Around ten o'clock I eat something, and drink a cup
(of) coffee. In the morning I like my coffee, and this
morning I drink even two cups. And then the phone is
ringing. That is the Moyer woman again, that gossip.
She knows everything, and she calls me up and tells me
everything. I don't like that at all, so I stay on the
phone only an hour.

I work hard the whole morning, and soon (before
long) half a day is past.

The noon hour is already here, and soon lunch (midday meal) is ready. Then I sit here again and eat my
lunch. I look out the window, and the letter carrier
(mailman) is going by. When does that fellow eat anyhow? And he hasn't been at my house for days. Why
doesn't someone write to me? Well, I don't have to
write to anyone either.

Right after the noon hour I work again; there is always something to do. The phone rings once or twice
more. Why do they tell me everything? They know I
don't like to gab on the phone.

And so it goes on (continues) the entire afternoon.
Now it is going on three o'clock, and I am <u>still</u>

an sellre Zeit kumme die Kinner ball heem vun der Schul.
Dann kenne mer all ebbes esse.

Net lang schpeeder iss mei Mann deheem. Er schafft
waerdaags bis fimf Uhr, un oweds kummt er ball nooch de
fimfe heem. Er hot dann en groosser Abbeditt--un ich
aa--un so muss es Nachtesse faddich sei. Nochemol sitz
ich do am Disch un guck aus em Fenschder. Wuhie geht
die denn? Ferwas bleibt sell Weibsmensch net deheem?

Nooch em Sobber lest mei Mann die Zeiding, un geht
naus. Er gleicht die Owetluft, graad eb es duschber
watt. Un dann geht die Sunn unner, un der Owetschtann
kummt raus. Es iss nau wennich vor Nacht.

Eblang watt es gans dunkel. Es iss Nacht; der Moond
geht uff, un es iss ball Bettzeit. Mer esse noch en
bissel ebbes. En Nochber geht verbei mit seim Hund.
En liewes Ding.

Noch en Daag verbei. Un do schtehn ich im Nachthemm; mer gehne ins Bett. Waardt en Minutt! Ich muss
noch ee mol gucke. Ya, datt isser (iss er). Darrich
die Nacht kann nix letz geh. Der Nachthaffe iss unnich
em Bett.

Vocabulary

der Abbeditt': appetite
    als: be in the habit of, "used to"
    als noch: still, yet
der Alt: old man
    babble, gebabbelt: to talk; chat, gossip
der Barrick, Barrigge: mountain
die Bettzeit: bedtime
der Breckfeschtdisch: breakfast table
der Brief, Briefe: letter
der Briefdreeger: letter carrier, mailman
die Daag(es)helling: dawn
    dernoh': after that
 es Ding, Dinger: thing
    drinke: to drink
    driwwe: over (there); datt driwwe: over there
    drowwe: up (there), above
    duschber: dusk
    eblang': before long

working. But at that time the children are soon coming
home from school. Then we can all eat something.

Not long after my husband is home. He works on
weekdays till five o'clock and in the evenings he comes
home soon after five. Then he has a big appetite--and
I do too--and so supper must be ready. Again (once
more) I am sitting here at the table and looking out
the window. Where is <u>she</u> going? Why doesn't she stay
at home?

After supper my husband reads the paper and goes outside. He likes the evening air, just before it gets dusk.
And then the sun goes down (under), and the evening
star comes out. It is now a little bit before night.

Before long it gets very dark. It is night; the
moon rises, and it is soon bedtime. We eat a little
bit something. A neighbor goes by with his dog. A
dear thing.

Another day past (gone). And here I stand in my
nightshirt (nightgown); we're going to bed. Wait a
minute! I must look once more. Yes, there it is.
Nothing can go wrong during the night. The chamber
pot is under the bed.

     ennichebber: anybody
     faddich: finished, ready
     ferwas?: why?
     finschder: dark
die Foohn: (tele)phone
     glei: immediately, right away
     glingele: to ring
     glitzerich: glittery
     halb: half
     heem: home
der Hund, Hund(e): dog
     kaafe: to buy
der Kaffi: coffee
 es Koppche, Koppcher: cup
     lieb: dear
 es Marriyedsesse: breakfast
der Marriyschtann: morning star
der Meelmann, -menner: mailman

```
 es Middaag(esse):  noonday meal
die Middaagsschtunn:  noon hour
die Millich:  milk
die Millichboddel, -boddle:  milk bottle
der Millichmann, -menner:  milkman
der Mo(o)nd:  moon
die Moyern:  (see note)
 es Nachtesse:  supper
 es Nachthemm, -hemmer:  nightshirt, nightgown
    neegscht:  next
    nei:  new;  ebbes Neies:  something new
der Nochber, Nochbere:  neighbor
    noh:  then
die Owetluft:  evening air
der Owetschtann:  evening star
    rabble:  rattle
    raus:  out
die Retsch:  gossip
    rufe:  to call
der Schtraahl, Schtraahle:  ray, beam
 es Sobber:  supper
die Sunn:  sun
    uffgeh:  to go up, "rise"
    unnergeh:  to go down, "set"
    unnich:  under
    verbei':  over, past
    verbei'geh:  to go past, to pass
    voll:  full
    wacker:  awake;  wacker warre:  to get awake
    waerdaags:  weekdays
die Weil:  while;  en Weil zrick:  a while ago
    wuhie?:  where to?
die Zeiding, Zeidinge:  newspaper
    z(e)rick:  back
```

Vocabulary Notes

drinke, kaafe, ringe, rufe: conjugated like kumme.

babble, gling(e)le, rabble: conjugated like rechle.

als: In this chapter, als is used as a particle meaning the action of the verb is carried out several
   times, many times, or even habitually. Ich geh als
   ans Fenschder "I go to the window several times,"

"I keep going to the window," "I'm in the habit of
going to the window." With past forms, als means
"used to": Ich waar als en Briefdreeger im Schteddel "I used to be a letter carrier in town."

als noch: These two words, treated as one unit, mean
"still" or "yet"; als must be very strongly accented.

duschber: also duschder.

ferwas': also warum'.

es Koppche: also Koppli, Kopplicher (Kopplin).

es Marriyedsesse: also es Marriye-esse. Es Breckfescht is a common E loan word.

die Moyern: The idea of adding an -n to a masculine
noun to form a feminine (der Tietscher, die Tietschern; der Meeschder, die Meeschdern) carries over
to last names if the name ends in -r or -l: der
Yoder, die Yodern; der Moyer, die Moyern; der Men-
gel, die Mengeln. If the name ends in a consonant,
one adds the ending -en: der Hendricks, die Hen-
dricksen. (One also finds -in in PG literature, re-
flecting the standard German ending.)

Grammar

A. The Demonstrative seller "that," "that one"

	Singular			Plural
	M.	F.	N.	All Genders
N.& A.	seller	selli	sell	selli
D.	sellem,sem	sellre	sellem,sem	selle

Like the demonstrative daer, seller can be used as
both adjective and pronoun, and like daer, any attributive pronouns that follow seller must have weak endings: seller alt Mann, sellem alde Mann.

Also, much as PG speakers use do after the demonstrative daer, just so they also use datt with seller,
but after the noun: seller Mann datt, selli Fraa datt.

B. The Interrogative <u>weller</u> "which," "which one"

We would be remiss if we did not immediately look at the interrogative <u>weller</u>, for as you will notice, it is exactly like <u>seller</u>:

	Singular			Plural
	M.	F.	N.	All Genders
N.& A.	weller	welli	well	welli
D.	wellem	wellre	wellem	welle

<u>Weller</u>, like <u>seller</u>, may be used as both adjective and pronoun:

Wellre Fraa gebscht es Buch?	To which woman are you giving the book?
Un wellre gebscht der Bensil?	And to which (one) are you giving the pencil?

CHAPTER FIVE

Part One

This lesson covers the months, seasons, and weather.
A few holidays during the year are thrown in for good
measure.

Nau hen mer die neie Kallener
Un sin aa schunn widder im Yenner;
Den hasse die Alde,
Yuscht weege dem Kalde.
Sie meene der Summer waer schenner.[1]

Im Yuli kummt mer an die Aern,
Der Bauer schwitzt, doch schafft er gaern;
Der Weeze schteht wie Gold im Feld
Un hungrich iss die gansi Welt.[2]

September watt gebluugt, ge-eegt,
Noh kummt mer an die Soot;
Do sehnt mer wie der Bauer schafft
Un sarrigt fers deeglich Brot.[3]

Disember uff der Bauerei,
Do kann's yo recht gemietlich sei;
Mer hot aa gschafft bei aller Hitz,
Nau sucht mer sich en weecher Sitz.[4]

John Birmelin, from [1]*"Der Yenner,"* [2]*"Der Yuli,"* [3]*"Der*
September," [4]*"Der Disember,"* *Later Poems.*

93

You may want to study the PG names of the months before you start this part of Chapter Five; they are all masculine gender (definite article <u>der</u>).

Yenner	Yuli
Hanning	Augscht (or Auguscht)
Maz	September
Abrill'	Oktower
Moi	Nowember
Yuni	Disember

<u>Tschuun</u> and <u>Tschulei</u> are very popular E loan words used by many PG speakers.

Die Munede

Es ist Halbnacht. Der eenundreissichscht Disember? Odder der aerscht Yenner? 'S macht nix aus. 'S iss Neiyaahr un mer kann schiesse heere. Die Neiyaahrschitz gehne Neiyaahr schiesse un winsche. Sie gehne rum zu ihre gude Freinde in der Nochberschaft, saage en Neiyaahrswunsch, un schiesse ihre Flinde in die Luft. Dann darrefe sie enwennich ins Haus geh un ebbes drinke un esse.

Noh winsche sie yederm "En glicklich Neiyaahr" un gehne weider. Fer ihre bessere Freinde sin die Winsche schee lang, awwer sie bhalde die lengschde un schennschde Winsche fer ihre beschde Freinde. Un sie bleiwe aa lenger ass (wie) bei de annere. Un drinke mehner. Un am Neiyaahrsdaag hen sie der gans Daag Koppweh, un verleicht fiehle sie net gut. Awwer am zwedde odder dridde geht alles besser.

Un bis der zwett Hanning, der Grundsaudaag, sin sie widder in beschder Adder. Noh kenne sie naus fer der glee Wedderprofeet, die Grundsau, waarde. Der wievielt iss heit? Ya, es iss der zwett Hanning, un des meent, die Grundsau kummt aus ihrem Loch un gebt uns ihre Wedderbericht. Sehnt sie ihre Schadde?

Am zwelfde Hanning feire mer em Lincoln sei Gebottsdaag, am vazehde iss es Valentinsdaag (duscht deim Bo en scheeni Kaart schicke?) un am achtzehde iss es em Washington sei Daag. (Awwer der zweeunzwansichscht

Watch also for comparative and superlative forms of adjectives, and three more modal auxiliaries: <u>darrefe</u> "to be allowed to," <u>meege</u> "may," and <u>wolle</u> "to want to."

January	July
February	August
March	September
April	October
May	November
June	December

The Months

It is midnight. The thirty-first of December? Or the first of January? It doesn't matter. It is New Year and one can hear shooting. The New Year shooters are "shooting in" the new year and going "wishing." They go around to their good friends in the neighborhood, say a New Year's wish, and shoot their guns into the air. Then they may go into the house a little (while) and drink and eat something.

Then they wish everyone a "Happy New Year" and continue on. For their better friends the wishes are nice and long, but they keep the longest and nicest wishes for their best friends. And they also stay longer than at the others. And drink more. And on New Year's Day they have a headache the whole day, and maybe they don't feel good. But on the second or third everything goes better.

And till the second of February, Groundhog Day, they are again in best order. Then they can go out and wait for the little weather prophet, the groundhog. What's the date today? Yes, it is February 2, and that means the groundhog comes out of its hole and gives us its weather report. Does she see her shadow?

On the twelfth of February we celebrate Lincoln's birthday, and on the fourteenth it is St. Valentine's Day (are you sending your Beau a nice card?), and on the eighteenth it is Washington's Day. (But February

Hanning iss sei Gebottsdaag.) Ee Daag schpeeder, am
neinzehde, iss es Faasenacht; dann mache un esse viel
Leit Faasenachtkichelcher, odder Fettkuche. Un am
zwansichschde Hanning iss es Eschemittwoch. Die Faschtzeit iss do.

Im Maz bassiert net viel; am siwwezehde iss es em
Seent Paedrick sei Daag, un am dreissichschde iss der
Palmsunndaag. Awwer der Wind bloost schtarrick im Maz,
un die Kinner losse ihre Keits hooch in die Luft
fliege. Sie gehne als heecher un als weider. Der Bu
mit der bescht Keit un der menscht Schnur grickt sei
Keit es heechscht un es weidscht. Am eenunzwansichschde iss es Friehyaahr do. Es watt ball waarm (warrem).

Wann kann en Mensch en Kalb sei? Am aerschde Abrill!
An sellem Daag kann er en Abrillekalb warre. (Guck!
Dei Schuhbendel is uff!) Un des Yaahr iss Oschdersunn-
daag am sexde Abrill. Dann darrefe die Kinner ihre
Oschderoier suche. Un die Weibsleit kenne ihre Osch-
derhiet, die Oschderbannets, waere.

In Moi feiere mer Muddersdaag am elfde un Gedechtnissdaag am sexunzwansichschde. Im Yuni hen die Vedder
ihre Daag, der Vaddersdaag, am fimfzehde. Nau iss aa
die Schul aus, un die Kinner wolle der gans Daag
schpiele. Des kenne sie net immer; sie misse aa ebbes
schaffe.

Awwer am vierde Yuli wolle sie nadierlich daer
greeschde Feierdaag im Summer feire, der Viert Tschulei.
Sie meechde dann gaern mit Feierwaerricks schpiele,
awwer sie darrefe die menscht Zeit net; ihre Eldre
losse sie net mit denne Dinger schpiele. Nau watt's
heess (der Summer waar schunn am eenunzwansichschde
Yuni do) un die Kinner gehne schwimme so oft (ass) sie
kenne. Yedes Kind gleicht sell.

Nau kummt der Aug(u)scht. Hoscht im Augscht Gebottsdaag? Wann net, dann hoscht im Augscht ken Feierdaag, sell iss gewiss. Dann kummt der September, un
die Kinner misse widder zrick in die Schul. Un am
dreiunzwansichschde iss es Harrebscht, odder Schpotyaahr; es Laab, die Bledder, vun de Beem griege nau
scheene Farrewe.

22 is his birthday.) One day later, on the 19th, it is
Shrove Tuesday; then many people make and eat "Fass-
nachts," or doughnuts. And on the 20th of February it
is Ash Wednesday. Lent is here.

Nothing much happens in March; St. Patrick's Day is
on the seventeenth, and on the thirtieth is Palm Sunday. But the wind blows hard in March, and the children let their kites fly high in the air. They go
higher and higher and farther and farther. The boy
with the best kite and the most string gets his kite
the highest and the farthest. On the twenty-first
Spring is here. It will soon get warm.

When can a person (human being) be a calf? On April
1st! On that day he can become an April calf (fool).
(Look! Your shoelace is untied!) And this year Easter
Sunday is on April 6. Then the children may look for
their Easter eggs. And the women can wear their Easter
hats, the Easter bonnets.

In May we celebrate Mother's Day on the 11th and Memorial Day on the 26th. In June the fathers have their
day, Father's Day, on the 15th. Now school is also out
and the children want to play the whole day. They
can't do that always; they must also do something
(work).

But on July 4 they naturally want to celebrate the
greatest holiday of the summer, the Fourth of July.
They would then like to play with fire-crackers, but
they may not most of the time; their parents won't let
them play with those things. Now it gets hot (summer
was already here on the 21st of June) and the children
go swimming as often as they can. Every child likes
that.

Now comes August. Do you have a birthday in August?
If not, then you have no holiday in August, that's for
sure. Then comes September, and the children must (go)
back to school again. And on the 23rd it is Fall or
Fall (synonyms); the foliage, the leaves of the trees
are getting nice colors now.

Am zwelfde Oktower feire mer em Columbus sei Daag.
Un zwee Woche schpeeder darrefe die Kinner alle Sadde
Gleeder un Falschgsichder aaduh, un sie gehne schpuucke
vun Haus zu Haus darrich die Nochberschaft un griege
Kaendi un Obscht vun de Nochbere. Der neegscht Daag,
der aerscht Nowember, iss dann der Allerheilichedaag.
Un des Yaahr hen mer aa en Leckschun Daag, odder Wahl-
daag, am vierde.

Un viel Mannsleit gleiche selli Zeit vum Yaahr; sie
nemme ihre Schrotflint, gehne yaage, un schiesse Haase
un Fassande. Awwer yeder Ameerigaaner gleicht der
greescht Feierdaag im Munet--der Beeddaag, odder es
Dankfescht (Dankdaag). An sellem Daag, dem siwweun-
zwansichschde Nowember, esst mer der Tarreki; in Deitch
heesst er der Welschhaahne odder es Welschhinkel. En
bissel mehner ass en Woch schpeeder iss der Grisch(t)-
munet, der Disember, do.

Nau watt's kalt; am zweeunzwansichschde Disember iss
der Winder do. Awwer des baddert niemand; die liebsch-
de Feierdaage, Die Weihnachde, un es liebscht Fescht,
der Grisch(t)daag, kumme am fimfunzwansichschde. So an
de Feierdaage rum kaaft die Familye en Beindbaam, en
Grischbaam, un schtelle en uff im Haus, abbaddich in
der Wuhnschtubb. (Odder der Paepp hackt eener runner
im Busch.) Der Belsnickel, odder Sandi Klaas, kummt
mit seim Sack un legt en Grischkindel fer yedes gut
Kind unnich der Grischbaam. Er bringt Gschenke wie
Schpielsache, un die Kinner hen dann am Grischdaag en
gudi Zeit. Yedem Kind sei Gschenk iss es bescht.

Un ee kazi Woch schpeeder iss es Yaahr widder rum.
Neinzeh hunnert achtzich iss verbei. Hallich Neiyaahr!

Vocabulary

die Adder: order; in Adder: in order
der Allerheilichedaag: All Saints Day
der Ameerigaan'er: American
    baddere: to bother
    bassier'e: to happen
der Beeddaag: Thanksgiving
der Beindbaam, -beem: pine tree
der Belsnickel: Santa Claus
 es Blatt, Bledder: leaf, page

On October 12 we celebrate Columbus's Day. And two
weeks later the children may put on all sorts of
clothes and false faces, and go spooking from house to
house through the neighborhood and get candy and fruit
from the neighbors. The next day, the first of November, is then All Saints Day. And this year we also
have an Election Day, or Voting Day, on the fourth.

And many men like that time of year; they take their
shotguns, go hunting, and shoot rabbits and pheasants.
But every American likes the greatest holiday in the
month--Thanksgiving, or Feast of Thanksgiving (Thank
Day). On that day, November 27th, one eats turkey; in
German it is called turkey gobbler or turkey hen. A
little more than a week later, Christmas month, December, is here.

Now it gets cold; on the 22nd of December Winter is
here. But that doesn't bother anyone; the dearest
holidays, Christmas, and the dearest feast, Christmas
Day, come on the twenty-fifth. Around the holidays the
family buys a pine tree, a Christmas tree, and sets it
up in the house, especially in the living room. (Or
Pop chops one down in the woods.) St. Nicholas, or
Santa Claus, comes with his sack and lays a Christmas
present for every good child under the Christmas tree.
He brings gifts like toys, and the children then have a
good time on Christmas Day. Every child's gift is the
best.

And one short week later the year is over again.
1980 is over. Happy (glorious) New Year!

der Busch, Bisch: woods, forest
der Dankdaag: Thanksgiving
 es Dankfescht: Thanksgiving
der Eschemittwoch: Ash Wednesday
    esse: to eat
die Faasenacht: Shrove Tuesday
    falsch: false, wrong
 es Falschgsicht, -gsichter: false face, mask
die Farreb, Farrewe: color
die Faschtzeit: Lent

der Fassant', Fassan'de: pheasant
der Feierdaag, -daage: holiday
   feiere (feire): to celebrate
die Feierwaerricks: fireworks
 es Fescht, Feschde: feast, festival
der Fettkuche: doughnut
   fliege: to fly
die Flint, Flinde: gun
der Gebotts'daag, -e: birthday
der Gedecht'nissdaag: Memorial Day
die Gleeder: clothes
der Grischtbaam, -beem: Christmas tree
der Grischtdaag: Christmas Day
 es Grischtkindel: Christmas present, gift
der Grischtmunet: the Christmas month (December)
die Grundsau, -sei: groundhog
der Grundsaudaag: Groundhog Day
 es Gschenk, Gschenke: present, gift
 es Gsicht, Gsichter: face
der Haas, Haase: rabbit
   hacke: to chop (hack)
   hoch: high
die Kaart, Kaarde: card
 es Kalb, Kelwer: calf
   kalt: cold
   kaz: short
die Keit, Keits: kite
 es Kichelche, Kichelcher: little cake
 es Koppweh: headache
 es Laab: foliage
die Leckschendaag: election day
 es Loch, Lecher: hole
   losse: to let, allow
die Luft: air
   meege: to care to, may; meechde: would care to
   (see note)
der Muddersdaag: Mother's Day
der Munet, Munede: month
 es Obscht: fruit
 es Oschderbannet, -bannets: Easter bonnet
die Oschdere (pl.): Easter
der Oschderhut, -hiet: Easter hat
der Oschdersunndaag: Easter Sunday
der Palmsunndaag: Palm Sunday

's: (see note)
der Sack, Seck: sack, bag
die Satt, Sadde: sort, kind
der Schadde: shadow
    schicke: to send
    schiesse: to shoot
der Schitz: shooter, gunner
die Schnur: string
 es Schpielsach, -sache: toy(s)
 es Schpuuck, -e: ghost, spook
die Schrottflint, -flinde: shotgun
    schtarrick: strong, (fast)
der Schuhbendel: shoe lace
    schwimme: to swim
    suche: to look for, seek
der Tarreki: turkey
der Vaddersdaag: Father's Day
der Valentinsdaag: St. Valentine's Day
    waarm (warrem): warm
    waere: to wear
der Wahldaag: election day
 es Wedderbericht, -berichde: weather report
der Wedderprofeet', -profeed'e: weather prophet
die Weihnachde (pl.): Christmas
der Welschhaahne: turkey (m.)
 es Welschhinkel: turkey hen
der Wind: wind
    winsche: to wish
die Wohnschtubb, -schtubbe: living room
    wolle (welle): to want to
der Wunsch, Winsche: wish
    yaage: to hunt, chase
    yeder: every(one)

Vocabulary Notes

baddere, bassiere, feiere, hacke, heere, kaafe, schicke,
    schtelle, schwimme, suche, and waere: conjugated
    like kumme.

esse, losse, schiesse, and winsche: conjugated like
    schwetze.

der Allerheilichedaag: also Allerheil.

es Blatt: also es Blaat.

Eschemittwoch: also Aschermittwoch.

die Faasenacht: also Fassnacht, Faasnacht, Faschtnacht.

die Faschtzeit: also die Faschde (pl.).

der Grundsaudaag: also der Dachsdaag.

es Gschenk: a popular E loan word is es Bresent, Bre-
  sents.

meechde: The subjunctive of meege, meecht, is used as
  much as, and perhaps even more than, meege. It is
  the equivalent of E "would like": Er meecht ins
  Schteddel "He would like to go into town." A complete explanation of the subjunctive mood will be
  presented in a future lesson.

's: Very often the neuter article es is pronounced as
  nothing more than an s sound preceding the neuter
  noun: 'S Kind schteht do "The child is standing
  here." 'S Kind sounds almost exactly like E
  "skinned."

schicke: Like gewwe, whatever you send is an accusative object (direct object), to whomever you send it
  appears as a dative object (indirect object): Er
  schickt seinre Mudder en Brief "He sends his mother
  a letter"; Sie schickt ihm (em) es Geld "She sends
  him the money."

winsche: To whom you make a wish, dative case; what
  you wish to someone, accusative case: Ich winsch
  dihr (der) en glickliches Neiyaahr "I wish you a
  happy New Year."
     On New Year's night, one may suddenly hear some-
  one (der Winscher) shouting a New Year's greeting
  (der Neiyaahrsgruss, der Neiyaahrswunsch). At the
  end of the greeting, a greeter shoots off his gun.
  The greeters are then often invited in for something
  to eat and drink.

Grammar

A. The Indefinite Adjective and Pronoun, yeder, "every-
   (one)."

As an indefinite adjective, <u>yeder</u> usually has the following forms:

	M.	F.	N.
N.& A.	yeder	yedi	yedes
D.	yedem	yedre	yedem

As an indefinite pronoun, <u>yeder</u> often has the following forms:

	M.	F.	N.
N.& A.	yederer	yederi	yedres
D.	yederm	yederer	yederm

Both E every(one) and PG <u>yeder</u> have singular forms only. The following examples help to show the uses of the adjective and pronoun <u>yeder</u>.

Lescht du yedes Buch do?	Are you reading every book here?
Ya, ich les yedres (yedes).	Yes, I'm reading every one.
Gebscht du yedem Kind en Zent?	Are you giving every child a penny?
Ya, ich geb yederm (yedem) es Geld.	Yes, I'm giving every one the money.

B. Adjectives--Comparative and Superlative Forms

1. In E, the ending -er is normally added to the positive of an adjective to form the comparative: small, smaller; great, greater; (but large, larger). The ending -est is normally added to the positive to form the superlative: small, smallest; great, greatest; (but large, largest). In PG, the comparative ending is normally -<u>er</u>, and the superlative ending is normally -<u>scht</u>. All of the following "normal" adjectives have appeared in Lessons One through Five:

Positive	Comparative	Superlative
frieh, early	frieher	es friehscht
nei, new	neier	es neischt
dumm, stupid	dummer	es dummscht
schnell, fast	schneller	es schnellscht
wennich, few	wennicher	es wennichscht

2. The following exceptions have all appeared in previous lessons:

b becomes w before adding -er:

    lieb, dear   liewer   es liebscht

t becomes d before adding any endings:

    laut, loud   lauder   es laudscht
    gscheit, smart   gscheider   es gscheidscht

one s is dropped before adding -scht to a positive ending in ss:

    weiss, white   weisser   es weisscht
    heess, hot   heesser   es heesscht

an e before l or r may be dropped before adding -er:

    deier, expensive deirer   es deierscht
    dunkel, dark   dunkler   es dunkelscht

3. Some PG adjectives not only add endings to form their comparatives and superlatives, but also "mutate" their stem vowels. Again, the following examples are gleaned from the first five lessons.

a	e	e
alt, old	elder	es eldscht
kalt, cold	kelder	es keldscht
lang, long	lenger	es lengscht

aa(h)	ee(h)	ee(h)
waarem, warm	weeremer	es weeremscht

o	ee	ee
schpot, late	schpeeder	es schpeedscht

oo(h)	ee(h)	ee(h)
hoch, high	heecher	es heechscht
grooss, large, big	greesser	es greesscht

u	i	i
yung, young	yinger	es yingscht

4. We know that glee and schee add -n before an adjectival ending. An n is also added before the comparative and superlative endings. In addition, the long stem vowel of the positive is shortened in the comparative and superlative:

    glee, small   glenner   es glennscht
    schee, beautiful schenner   es schennscht

5. Two adjectives used in the first five lessons are irregular (as are, of course, their E equivalents):

gut, good   besser   bescht
viel, much,many   meh, mehner   menscht

Neither <u>meh</u> nor <u>mehner</u> take adjectival endings:

Hoscht meh Schnur?   Do you have more string?
Er hot mehner Schnur   He has more string than I.
ass ich.

6. However, adjectives normally take adjectival endings whether they appear in positive, comparative, or superlative forms, e.g., en alder Mann, em eldere Mann, me eldschde Mann; en yungi Fraa, re yingere Fraa, die yingscht Fraa. Don't let comparative and superlative endings fool you into thinking no adjectival endings are needed; simply add the correct adjectival endings <u>to</u> the comparative or superlative endings.

7. The E expression as + positive + as (e.g., as tall as) is expressed in PG with <u>so</u> + positive + <u>ass</u>: <u>Mei Bruder iss so grooss ass ich</u> "My brother is as big (tall) as I"; <u>ich</u> and "I" are subjects of the "understood" verb sei "to be."

8. The E expression comparative + than (taller than) is expressed in PG with comparative + <u>ass</u> (or <u>ass wie</u>): <u>Mei Bruder iss greesser ass (ass wie) ich</u> "My brother is taller than I."

9. The PG speaker normally precedes a superlative with a definite article or a possessive when the superlative is used as an attributive adjective: <u>Sie bhalde die lengschde Winsche fer ihre beschde Freinde</u> "They keep the longest wishes for their best friends."

But when used as a predicate adjective, superlatives regularly take no ending and are preceded by the neuter article <u>es</u>: <u>Yedem Kind sei Gschenk iss es bescht</u> "Every child's present is the best."

10. We have noted that adverbs do not take endings; this is true even when an adverb appears in its comparative or superlative form:

Sei Keit fliegt heecher
ass meini, awwer em
Tschanni sei Keit
fliegt es heechscht.

His kite flies higher
than mine, but Johnny's
kite flies the highest.

11. In E, comparative + and + comparative (higher
and higher, farther and farther) can be both adjective
and adverb. In PG this "double comparative" is expressed with als + comparative:

Mei eldschder Bu watt
als greesser.
Die Keits fliege als
heecher un als weider.

My oldest son is getting
taller and taller.
The kites are flying higher and higher, and farther
and farther.

C. Verbs

1. Modal Auxiliaries (Continued)

Chapter Four introduced us to the modal auxiliaries
kenne, solle, misse, and the modal-like verb brauche.
Let us now take a look at the modals darrefe "to be
allowed to"; meege "to care (wish) to, may": and wolle
"to want to."

2. darrefe (daufe, daerfe) "to be allowed to"

Singular	Plural
1. ich darref (dauf, daerf)	1. mir,mer darrefe (daufe, daerfe)
2. du darrefscht (daufscht,daerfscht)	2. dihr,der darreft (dauft); ihr,er dar-refe (daufe, daerfe); etc.
3. er,sie,es darref (dauf, daerf)	3. sie darrefe (daufe, daerfe)

3. meege "to care to," "may"

Singular	Plural
1. ich maag	1. mir,mer meege
2. du maagscht	2. dihr,der meege; ihr,er meecht; etc.
3. er,sie,es maag	3. sie meege

4. __wolle__ (__welle__) "to want to"

Singular	Plural
1. ich will	1. mir,mer wolle (welle)
2. du witt	2. dihr,der wolle (welle);
	ihr,er wott (wett); etc.
3. er,sie,es will	3. sie wolle (welle)

5. Whatever was said in Chapter Four about __kenne__, __misse__, and __solle__ can be said also about __darrefe__, __meege__, and __wolle__.

(a) An infinitive dependent on a modal goes to the end of a clause (sentence):

Er maag Muundaag oweds
Deitsch schwetze.

He cares to (may) speak
German on Monday evenings.

(b) An answer need not contain the dependent infinitive if it is obvious:

Witt (du) Deitsch
schwetze?
Ya, ich will.

Do you want to speak German?
Yes, I want to (Yes, I do).

(c) A dependent infinitive that indicates "going" need not be used at all:

Witt (du) ins Schteddel?

Do you want (to go) into town?

6. Double Infinitives

In Chapter Four we learned that __geh__ can be used with infinitives much as modals are used with infinitives:

Ich geh im Harrebscht
yaage.
Ich will im Harrebscht
yaage geh.

I am going hunting in the Fall.
I want to go hunting in the Fall.

Two other verbs used in __Die Munede__ can be used in just this same way: __heere__ "to hear" and __losse__ "to let, allow."

Ich heer die Schitz
schiesse.
Ich kann die Schitz
schiesse heere.

I hear the shooters (gunners) shooting.
I can hear the gunners shooting.

Die Kinner losse ihre
Keits hoch in die Luft
geh.

The children let their
kites go high into the
air.

Eldre solle ihre Kinner
net mit Feierkraeckers
schpiele losse.

Parents should not let
their children play with
firecrackers.

CHAPTER FIVE

Part Two

This portion of Lesson Five covers various expressions concerning the weather. You may wish to study the PG names of the seasons before you begin the reading selection.

es Friehyaahr   spring
der Summer   summer
es Schpotyaahr or der Harrebscht   autumn
der Winder   winter

Winder verbei,
Kummt Frieyaahr rei;
Die Aerd weckt uff
Un alles druff.[1]

Was iss so hibsch wie'n Summerdaag
  Mit Graas un griene Beem?
Die Sunn so waarm, der Wind so sanft,
  Un alles so genehm.[2]

Im Harrebscht so iss ken schenn're Zeit--
  Weescht net wie gut ich fiehl,
Wann Daage alsnoch schee sin, un
  Die Nachde bissel kiehl.[3]

Ich gleich en net, der Winder kalt,
  Un aa net Eis un Schnee;
Ich saag dir was es bescht mir gfallt--
  Sell iss's Friehyaahr schee.[4]

Ralph Funk, from [1]*"Friehyaahr,"* [2]*"En Summerdaag,"* [3]*"Im Harrebscht,"* [4]*"Der Winder Kummt,"* Poems of Ralph Funk.

Es Wedder

Heerscht ebbes? Ebbes wie viel Hunde? Mer kann sie blaffe (gauze) heere. Sie kumme als neecher un warre als lauder. Waard en Minutt! Sell sin yo gaar ken Hunde. Sell sin Gens! Guck mol gans datt drowwe im Himmel. Sie fliege gegge Nadde. Nau wisse mer fer schuur; es iss Friehyaahr.

Es Friehyaahr waar schunn am eenunzwansichschde Maz do, awwer es waar noch kalt and es Wedder waar noch wiescht. (Am zwette Hanning waar es gewiss sunnich!) Der Maz iss nau beinaah (fascht) verbei. Es iss immer noch windich, awwer ball sehne mer die Amschel.

Die Daage warre als lenger, un die Sunn gebt uns mehner Hitz un lenger Licht ass wie im Winder. Un im Abrill fallt der Regge. Hoscht dei Scheerm (Umberell)? Des iss yo en Suddelwedder. Awwer nau hen mer mehner scheene, waarme Friehyaahrsdaage, un ennihau, en nasser Abrill helft de Moiblumme waxe. Dann griege die Beem Bledder, un die Ebbelbeem bliehe schee weiss.

Der eenunzwansichscht Yuni iss der lengscht Daag im Yaahr. Es iss Summer. Im Tschun kann es marriyeds kiehl sei, awwer nammidaags watt's schee waarm. Un im Yuli iss es marriyeds schunn weermer, un nammidaags watt's recht heess--un als heesser. Un es kann arrig schwiel warre wann ken Liftel geht. Dann schwitzt mer vun der Hitz, un mer muss die Schweesdrobbe abwische. Do kann mer yo drauss uff em Feld en Sunneschtich griege. Heess Wedder gleiche viel Leit net.

Im Summer gehne die Kinner gaern Baarfiessich. Un des darrefe sie aa. Ihre Eldre losse sie unne Schuh rumschpringe. Awwer sie wolle liewer im alde Schwimm-loch schwimme geh. Sie meechte gude Schwimmer warre.

Alsemol im Summer watt's gans wolkich; die Sunn geht hinnich dunkle Gewidderwolke, un mer kann es dunnere heere. Dann heert mer der Dunner un sehnt der Blitz als neecher kumme. Uff eemol kumme die groosse Regge-droppe; es reggert--en rechder Wolkebruch. Nau iss der Gewidderschtarrem graad iwwer eem, un der Blitz un der Dunner sin graad wie eens.

Es wedderleecht (blitzt) so fer en halwi Schtunn, un mer muss hoffe, es dutt net aryeds (aeryeds) im Haus

The Weather

Hear something? Something like many dogs? One can
hear them barking. They are coming ever closer and
getting ever louder. Wait a minute! Those aren't dogs
at all. They are geese. Just look way up there in the
sky. They are flying towards the north. Now we know
for sure: it is Spring.

Spring was already here on the twenty-first of March,
but it was still cold and the weather was still ugly.
(On the second of February it was certainly sunny!)
March is now almost over. It is still windy, but soon
we'll see the robin.

The days get longer and longer, and the sun gives us
more heat and longer light than in the winter. And in
April the rain falls. Do you have your umbrella? This
is surely drizzly weather. But now we have more nice,
warm Spring days, and anyhow, a wet April helps the May
flowers grow. Then the trees get leaves, and the apple
trees bloom (blossom) nice and white.

The twenty-first of June is the longest day in the
year. It is summer. In June it can be cool in the mornings, but in the afternoons it gets nice and warm. And
in July it is already warmer in the mornings and in the
afternoons it gets right hot--and even hotter. And it
can get very sultry when there is no breeze. Then one
sweats from the heat, and one has to wipe off the beads
(drops) of sweat. One can actually get a sunstroke out
in the field. Many people do not like hot weather.

In summer the children like to go barefoot. And they
may do that. Their parents let them run around without
shoes. But they rather go swimming in the old swimming
hole. They would like to become good swimmers.

Sometimes in summer it gets very cloudy; the sun goes
behind dark storm clouds, and one can hear it thunder.
Then one hears the thunder and sees the lightning come
ever closer. Suddenly the big raindrops come. It rains
(is raining)--a veritable cloudburst. Now the thunderstorm is right over one and the lightning and the thunder are just like one.

It lightens thus for a half hour, and one has to hope
it doesn't strike (into) the house or the barn. But

odder in der Scheier eischlaage. Awwer deswegge hot
mer lange Gewidderrude uff em Dach vum Haus, un sogaar
lengere uff em heechere Scheierdach. Viel Leit farrichde (faerrichde) so en Schtarrem. Sie liewe es net,
wann es weddert; lauder Dunner un heller Blitz gleiche
sie gaar net.

Nau iss es Gewidder verbei, un es iss am Abhelle.
Gans weit fatt kann mer es wedderleeche sehne un dimmle
heere. Un nau kummt die Sunn raus, un mer sehnt en
groosser Reggeboge mit seine scheene Farrewe.

Mannichmol kann es im Summer, abbaddich im Tschulei
un Augscht, arrig drucke warre. Wochelang fallt ken
Regge, un die Felder warre darr (daerr). Druckner
Grund misse die Bauere beinaah mehner farrichde ass es
Gewidder. Do will nie nix richtich waxe.

Am dreiunzwansichschde September iss der Harrebscht
(es Schpotyaahr) do. Die Luft watt nau kiehler, un die
Daage warre als kazer. Die Gens fliege gegge Sudde.
Awwer im Harrebscht watt es mannichmol widder summerich.
Der Aldweiwersummer kummt, un yederebber iss froh. Mer
hen widder waarme Marriye un fascht heesse Nammidaage.

Eblang sehnt mer der Reife (der Froscht) uff em
Grund un uff de Decher un ball will die Zeiding Schnee
hawwe. Mannichmol iss die Luft immer noch en bissel zu
waarm, dann dutt's net schnee-e, awwer es kisselt odder
haggelt. Gleenes, weisses Kissel odder der greesser
Haggel un die greessere Schloosse hupse rum wie Ping-
Pong Balle. Des iss noch en Hunswedder.

Un wann so en Kisselwedder odder Schloossewedder
kummt, iss der Boddem wunnerbaar schlipperich. Mer
kann schlippe, odder ausrutsche. Awwer hoch drowwe uff
em Barrig sehnt mer schunn weisser Schnee. Net lang
dennoh falle die aerschde Schneeflocke aa im Daal.
Hoscht dei Schneeschipp (Schneeschauffel) hendich?

Am zweeunzwansichschde Disember iss der Winder do.
Es iss aa der kazschde Daag im Yaahr. Nau warre die
Daage lenger, awwer net weermer. Im Yenner saagt mer,
wie lenger die Daage warre, wie kelder watt's Wedder.
Un sell iss gewiss un waahr.

Es iss Hossesackwedder; nau kumme die keldschde Win-
derdaage un die greeschde Schneeschtarreme, un der Winderwind, der Naddwind, friert es Wasser im Deich. Der

that's why one has the long lightning rods on the roof
of the house and even longer ones on the higher barn
roof. Many people fear such a storm. They don't like
(it) when it storms; loud thunder and bright lightning
they don't like at all.

Now the thunder storm is past, and it is clearing
off. Very far away one can see it lighten and hear
distant thunder. And now the sun comes out, and one
sees a big rainbow with its (his) beautiful colors.

Sometimes in summer, especially in July and August,
it can become very dry. For weeks no rain falls, and
the fields get dried out. The farmers must fear dry
ground (earth) almost more than the thunderstorm.
Nothing ever wants to grow right.

On the twenty-third of September Fall is here. The
air gets cooler now, and the days get shorter and shorter. The geese fly south. But in the Fall, it sometimes gets summery again. The Indian Summer is coming,
and everyone is happy. We have warm evenings and almost hot afternoons.

Before long one sees frost on the ground and on the
roofs, and soon the paper "wants" snow. Sometimes the
air is still a little too warm; then it doesn't snow,
but it sleets or hails. Little, white sleet or the
larger hail and the bigger hailstones jump around like
ping-pong balls. This is terrible weather.

And when such sleety weather or hail storms come,
the ground is very slippery. One can slip, or slip
(synonyms). But high above on the mountain one sees
already white snow. Not long thereafter, the first
snowflakes fall also in the valley. Do you have your
snow shovel handy?

On the twenty-second of December, Winter is here.
It is also the shortest day of the year. Now the days
get longer, but not warmer. In January one says, the
longer the days get, the colder the weather gets. And
that is certain and true.

It is hands-in-the-pockets weather; now come the
coldest winter days and the biggest snowstorms, and the
winter wind, the North wind, freezes the water in the

Sandy Klaas waar awwer gut zu de Kinner, un sie hen
neie Schlidde un Schkeets. Sie meinde die Kelt un der
Wind gaar net, un darrefe Schlidde faahre un schkeede,
un Schneeballe schmeisse. Der Schnee iss dief un es
Eis iss dick.

Es iss nau Maz. Der Winder iss ball verbei. Es
Schleffelwedder kummt; die keldschde Daage sin verbei,
un der Schnee un es Eis sin am Schleffle, odder Schmelse. Ball kumme die Gens widder.

Vocabulary

abhelle: to clear (as after a storm)
abwische: to wipe off
der Aldweiwersummer: Indian summer
die Amschel, Amschle: robin
aryeds (aeryeds): somewhere
ausrutsche: to slip
der Baam, Beem: tree
baarfiessich: barefooted
der Ball, Balle: ball
blaffe: to bark
bliehe: to bloom, blossom
der Blitz: lightning
blitze: to lighten, lightning
die Blumm, Blumme: flower
es Daal, Deller: valley
es Dach, Decher: roof
darr (daerr): dry
der Deich, Deiche: pond
dennoh: after that, thereafter
dick: thick, fat
dief: deep
dimmele: to thunder (in the distance)
der Drobbe: drop
drucke (druckn + ending): dry
der Dunner: thunder
dunnere: thunder
der Ebbelbaam, -beem: apple tree
es Eis: ice
eischlaage: to strike (lightning)
falle: to fall
farrichde (faerrichde): to fear
fascht: almost

pond. But Santa Claus was good to the children, and
they have new sleds and skates. They don't mind the
cold and the wind at all, and may go sledding and skating, and throw snowballs. The snow is deep and the
ice is thick.

It is now March. Winter is soon over. The thaw
weather is coming; the coldest days are over, and the
snow and ice are thawing, or melting. Soon the geese
will be coming again.

 es Feld, Felder: field
der Friehyaahrsdaag: spring day
    friere: to freeze
    froh: happy, glad
der Froscht: frost
die Gans, Gens: goose
    gauze: to bark
    gegge: toward (against); gegge Nadde: toward the
       north; gegge Sudde: toward the south
 es Gewidder: thunderstorm
die Gewidderrud: lightning rod
der Gewidderschtarrem: thunderstorm
die Gewidderwolk, -wolke: thundercloud
der Haggel: hail
    haggele: to hail
    heess: hot
    hendich: handy
die Hitz: heat
    hoffe: to hope
 es Hossesackwedder: (see notes)
 es Hunswedder: bad weather
    hupse: to hop
die Kelt: cold
    kiehl: cool
der Kissel: sleet
 es Kisselwedder: sleety weather
    kissle: to sleet
 es Licht, Lichder: light
 es Liftel: breeze
    manchmol: sometimes
    meinde: to mind, remember
die Moiblumm, -blumme: May flower
der Naddwind: north wind

der Regge: rain
der Reggeboge: rainbow
der Reggedrobbe: raindrop
    reggere: to rain
der Reife: frost
    richtich: right
    rumschpringe: to run around
der Scheerm: umbrella
die Scheier, Scheire: barn
 es Scheierdach, -decher: barn roof
    schkeede: to skate
die Schkeets (pl.): skates
 es Schleffelwedder: thawing weather
    schleffle: to thaw
der Schlidde: sled; Schlidde faahre: to go sledding
    schlippe: to slip
    schlipperich: slippery
die Schlooss, Schloosse: hailstone
 es Schloossewedder: hail storm
    schmeisse: to throw
    schmelse: to melt
der Schneeball, -balle: snowball
    schnee-e: to snow
der (die) Schneeflocke: snowflake
die Schneeschaufel, -schaufle: snow shovel
die Schneeschipp, -schibbe: snow shovel
der Schneeschtarrem: snowstorm
der Schtarrem, Schtarreme: storm
der Schuh: shoe
    schur: sure; fer schur: for sure
der Schweess: sweat
    schwiel: sultry
der Schwimmer: swimmer
 es Schwimmloch, -lecher: swimming hole
    schwitze: to sweat
 es Suddelwedder: drizzly weather
    summerich: summery
die Sunn: sun
der Sunneschtich: sunstroke
    sunnich: sunny
die Umberell: umbrella
    unne: under
    waahr: true
 es Wasser: water
    waxe: to grow

```
        weddere:  to weather (thunder and lightning)
        wedderleeche:  to lighten (lightning)
        weit:  far;  weit fatt:  far away
der Winderdaag:  winter's day
der Winderwind:  winter wind
        windich:  windy
        wochelang:  for weeks
die Wolk, Wolke:  cloud
die Yaahreszeit, -zeide:  season
```

Vocabulary Notes

abhelle, blaffe, bliehe, dunnere, falle, friere, hoffe,
    meinde, reggere, schlippe, schwimme are conjugated
    like kumme.

abwische, ausrutsche, blitze, gauze, hupse, schmeisse,
    schmelse, schwitze are conjugated like schwetze.

dimmele, if written and pronounced dimmle, kissle,
    schleffle are conjugated like rechle.

waxe: The second person singular of waxe is spelled du
    wackscht.

ab-, aus-, ei-, rum-, are separable prefixes that will
    be taken up formally in a future lesson.

der Altweiwersummer: also der Inschingsummer, der In-
    schesummer.

die Gewidderrud: also die Wedderrud.

es Hossesackwedder: Weather that is so cold that one
    has to keep hands in pockets in order to keep them
    warm.

manchmol: also mannichmol(l).

meinde: also has the meaning "to remember" for many PG
    speakers.

der Schneeflocke: also der Schneebrocke.

Grammar

A. Strong Adjectival Endings

    1. Both E and PG speakers do not always want to

precede an attributive adjective with a definite or indefinite article, or a possessive or demonstrative adjective, e.g.:

| I like black coffee. | Ich gleich schwazer Kaffi. |
| Good children get nice presents. | Gude Kinner griege scheene Bresents. |

When PG attributive adjectives are themselves not
preceded, they take "strong" endings.

The Strong Declension
Singular

	M.	F.	N.
N.&A.	alder Mann	scheeni Fraa	glee Kind
D.	aldem Mann	scheener Fraa	gleenem Kind

Plural
All Genders

| N.&A. | alde (scheene,gleene) Menner (Weiwer,Kinner) |
| D. | alde (scheene,gleene) Menner (Weiwer,Kinner) |

Months, weather, comparatives, and superlatives:

Im _____ watt's schunn _____; im _____
   (April)   (warm)   (May)

watt's _____; un im _____ watt's _____
    (warmer)   (June)   (warmer and

_____. Viel Daage in sellem Munet warre sogaar
warmer)

_____. Der _____ iss nadierlich viel _____,
(hot)   (August)   (hotter)

awwer der _____ iss gewiss _____. Im
      (July)   (the hottest)

_____ watt's _____, awwer net _____
(September)   (cool)   (as cool as)

im _____.
  (October)

Der _____ iss gewiss schunn _____, un
  (November)   (cold)

_____ iss nadierlich _____; awwer der _____
(December)   (colder)   (January)

un der _____ sin die _____ Munede im Yaahr.
  (February)   (coldest)

Die aerschde paar Woche im _____ sin immer noch
              (March)

_____, vielleicht _____ der _____;
(cold)   (as cold as)   (November)
awwer dann warre die Daage nau _____, un sie warre
                                 (longer)
_____ bis _____. Dann sin die
(longer and longer)   (June)
Daage _____, viel _____ im _____.
      (the longest)   (longer than)   (December)
Awwer der eenunzwansichscht _____ iss _____
                     (June)   (the longest)
Daag im Yaahr. Die Daage warre dann _____, bis am
                            (shorter)
zweeunzwansichschde _____, em _____ Daag vum
            (December)   (shortest)
Yaahr, etliche Schtunn _____ im _____.
           (shorter than)   (June)

CHAPTER SIX

Part One

Our young friend may be taking us through a specific
house--his own home--but the material in this lesson
lends itself to all homes. The vocabulary is almost
overwhelming, but pick and choose what you can use.
You can always come back and pick up the rest.

We have been reading sentences containing separable
prefixes since the very first chapter, but they were
disguised as mere particles, especially since they
matched their E counterparts so closely. In the following reading you will find in parentheses the infinitives of verbs with separable prefixes. An explanation
of prefixes will be found in the grammar section of
this lesson.

Also, this lesson will finally enlighten you as to
the cases used with the various prepositions.

Heem meent meh ass Land un Haus,
Schannschtee, Fenschdre, Drin un Draus,
Keller, Schpeicher, Bettschtubb, Dach,
Kich un Gaarde, Beem un Bach.

Haus un Hof, ken Heemet noch!
Muss mol zeitlich kumme doch
Menschelieb un Kinnerschpiel,
Lieb, viel Lieb un Lewe viel.[1]

[1]*Russell W. Gilbert, from "Heemet,"* Bilder un Gedanke.

Es Haus; Es Aerscht Deel

Meinre Familye ihre Heemet iss en zimmlich alt Haus ungefehr ee odder zwee Meil vum Schteddel. Es iss aus Backeschtee, en recht alt Backeschteehaus. Es hot zwee Schteck un en ewwerschder Schpeicher. Loss uns mol des Haus vun ausse aasehne.

Es hot en vedderschdi Bortsch so breet wie's Haus. Vanne kann mer aa nein Fenschdre sehne, vier im aerschde Schtock un fimf im zwedde Schtock. Es Haus hot ken Dachfenschdre. Die Fenschdre hen alle Leede, un mer kann sie zumache, wann es weddert. Es Dach iss en Schindeldach; die Schindle sin aus Holz, awwer sie sin gut un dick, un es Reggewasser kummt naeryeds ins Haus rei (reikumme), es kann schtarreme, wie es will. Mer kann aa zwee Schannschtee sehne, en Gewidderrud an eem Giwwelend, un en Kandel an yedere(r) Eck vum Haus.

En Fens schteht gschwischich em Haus un der Schtrooss. Sie geht rechts an sellem Schaddebaam datt verbei (verbeigeh) un dann an der Hauseck hinnenaus bis zum Hinnerhof. Die Schaddebeem waxe graad newwich em Haus. Sie warre yedes Yaahr greesser, un ihre Nescht waxe schunn wedder's Haus. Selli misse mer mol abhacke odder abseege.

Kumm, mer gehne mol ins Haus nei (neigeh). Aerscht misse mer die drei Bortschdreppe nuffgeh, un uff die vedderscht Bortsch geh. Do uff der Bortsch sin en paar Schtiehl, Benk un der Gremmemm ihre Schockelschtuhl. Mannchmol sitzt sie oweds drin un guckt zu (zugucke), wie die Sunn iss am Unnergeh.

Do iss unser Hausdier, die vedderscht Dier. Die ewwerscht Helft hot vier Fenschderscheiwe. Die losse es Daageslicht reikumme. Wu iss mei Hausschlissel? Unne en Schlissel grickt mer die Dier net uff (uffgriege). Mer schtecke der Schlissel ins Schlisselloch nei (neischtecke), drehe en rum (rumdrehe) un schliesse die Dier uff (uffschliesse).

Dann drehe mer der Diergnobb, mache die Dier uff (uffmache), un gehne ins Haus nei. Mer mache die Dier zu (zumache), un do schtehne mer im Vorgang vum aerschde Schtock.

Gans hinne sehne mer die Kich, awwer ich geh mol

The House; Part One

My family's home is a rather old house about one or
two miles from town. It is (made) of brick, a really
old brick house. It has two stories (floors) and an
attic. Let us (once) take a look at the house from the
outside.

It has a front porch as wide as the house. In front
one can see nine windows, four in the first floor and
five in the second floor. The house has no dormer windows. The windows all have shutters, and one can close
them when it storms. The roof is a shingle roof; the
singles are of wood, but they are good and thick, and
rain water doesn't come into the house anywhere, it can
storm as much as it wants to. One can also see two
chimneys, a lightning rod at one gable end, and a spout
at every corner of the house.

A fence stands between the house and the street. It
goes to the right past that shade tree there, and then
at the corner of the house out to the back yard. The
shade trees grow right next to the house. They get
larger every year, and their branches grow already
against the house. We must chop or saw them off some-
time.

Come, we'll go into the house. First we must go up
the three porch steps and onto the front porch. Here
on the porch are a few chairs, benches, and grandma's
rocking chair. Sometimes she sits in it in the evening
and watches how the sun sets (is going down).

Here is our house door, the front door. The top
half has four window panes. They let the daylight in.
Where is my house key? Without a key one doesn't get
the door open. We stick the key into the keyhole, turn
it, and unlock the door.

Then we turn the doorknob, open the door, and go into the house. We close the door, and here we stand in
the vestibule of the first floor.

All the way back we see the kitchen, but I am going

aerscht do nei. Kumm mit mer rei (reikumme). Des iss
die Wuhnschtubb, die schennscht Schtubb im Haus. Do
hot die Memm ihre beschde Sache--Soofe (Settie), un
Sessel, weisse Vorheng an de Fenschdre, un en Karrepet
uffem Boddem. Der grooss Feierheerd hot zwee Feierhund
un aa en Mantelbord. Summers schteht en scheeni Uhr
datt druff (druffschteh).

Bilder henge an de Wend (gleichscht du es Wandbabier?) un en Bicherschank schteht zwische de zwee
Fenschdre. Meinre Mudder ihre(r) Meening nooch soll
nimmand do reikumme--du weesscht, weeich (weege) de
Kinner. Ich hab nix degeege; ich bin sogaar defor.
Dann bleibt alles in der Schtubb immer in Adder un
schee sauwer.

Mer kenne nau widder in der Gang nausgeh, awwer mer
gehne liewer aerscht darrich die Dier do un in die Essschtubb nei. Mer muss do uffbasse; die Dierschwell iss
en bissel hoch. In der Essschtubb esse mer Sunndaags
un an (uff) Feierdaage. Mer kenne der Disch greesser-
mache, viel lengermache, un dann hole mer mehner
Schtiehl rei (reihole), un fascht die gans Freindschaft
kann dohin esse. Datt im Glassschank hot die Memm ihre
bescht Gscharr (Gschaerr). Drin in der Kich hen mer
noch annere Schenk.

Kumm, mer gehne nau in die Kich nei. Ya, ich weess,
die Memm iss am Koche, awwer kumm ennihau rei. Schee
grooss, gell? Do waar mol en alder, schwazer Holz- un
Kohleoffe datt in der Eck, awwer zidder 1955 hen mer en
weisser Lecktrickoffe. Uff em Offe druff kann mer
koche, un im Backoffe drin kann mer backe.

En lang(es) Offerohr waar als do graad unnich der
Deck. Mer brauche des nimmi. Awwer die alt Holskischt
iss immernoch newwich em neie Kochoffe. Mer duhne nadierlich ken Brennhols in die Kischt nei (neiduh), awwer unser Hund, der Wuffi, schlooft gaern hinnich der
Kischt, un im Winder hupst die Katz gaern nuff (nuffhupse) uff die Kischt un leit do druff (druffleige).

Datt im Eckschank, hinner denne gleene Diere, iss
mannicher Deller un Haffe, mannichi Schissel un Pann,
un mannich Glaas, Blettli, un Koppche (Koppli). Die
Deckel sin all unne. In denne Schubblaade do in dem
do groosse Kicheschank sin Messer (Dischmesser, Brot-
messer, sogaar greessere Butschermesser), Gawwele

first of all in here. Come in with me. This is the
living room, the nicest room in the house. Here Mom
has her best things--sofa (settee), and easy chairs,
white curtains on the windows, and a carpet on the
floor. The large fireplace has two andirons and also a
mantelpiece. During the summers a beautiful clock
stands on it.

Pictures hang on the walls (do you like the wallpaper?) and a book case stands between the two windows.
According to my mother's opinion, no one should come in
here--you know, because of the children. I have nothing against that; I am even for it. Then everything in
the room will stay in order and nice and clean.

We can go out into the hallway again but we better
(rather) go through this door and into the dining room.
One has to be careful (watch out) here; the door sill
is a little high. In the dining room we eat on Sundays
and on holidays. We can make the table larger, make it
much longer, and then we fetch more chairs in, and almost the whole family clan can eat in here. There in
the china cupboard Mom has her best dishes. In the
kitchen we have also other cupboards.

Come, we'll go into the kitchen now. Yes, I know,
Mom is cooking, but come in anyhow. Nice and big,
right? There was once an old, black wood and coal
stove in the corner, but since 1955 we have a white
electric stove. One can cook on top of the stove and
bake in the oven.

A long stovepipe used to be right under the ceiling.
We don't need it anymore. But the old wood box is
still next to (beside) the new stove. We naturally
don't put any more wood into the box, but our dog Wuffi
likes to sleep behind the box, and in winter the cat
likes to jump up onto the box and lie on it.

There in the corner cupboard, behind those little
doors, is many a plate and pot, many a bowl and pan,
and many a glass, saucer, and cup. The covers (lids)
are all below (underneath). In these drawers in this
big kitchen cupboard are knives (table knives, bread
knives, even larger butcher knives), forks (large and

(groosse un gleene Kiche- un Dischgawwele), un aa
Leffle (Essleffle, Tee- un Kaffileffle, Suppeleffle, un
Scheppleffle).

Datt owwe hot die Memm ihre Sals un Peffer, Zucker
un Kaffi, un ihre Gewaerz; un datt unne hot sie Pitschers, die Kaffikann, un der Teekessel. Die groosse
eisne Kessel hot sie in der Summerkich. Die do Dier
fiehrt in die Summerkich nei (neifiehre), un selli
Kichedier datt fiehrt naus (nausfiehre) in der Hof un
zum Gaarde. (Laaf awwer net darrich die Mickedier; die
muss mer aerscht uffmache!)

Do iss die Schpielbank. Mer dreht do, un's Wasser
kummt an dem do Graahne raus (rauskumme). Dann dreht
mer widder zu (zudrehe), un schpielt 's Gscharr
(Gschaerr) ab (abschpiele) in der Schpielschissel mit
em Schpiellumbe. Mer drickelt sie dann ab (abdrickle).
Do sin noch Dischlumbe un Handdicher.

Der Essschank iss datt hinnich sellre Dier. Unser
Esssache (Lewesmiddel) sin do drin. Un newwich em
Kichedisch schteht meim gleene Bruder sei Hochschtuhl
(Hochschtiehlche).

Ferwas iss der Kichedisch so grooss? Ich hab siwwe
Geschwischder!

Vocabulary

    aasehne: to look at
    abdrickle: to dry off
    abhacke: to chop off
    abschpiele: to rinse (wash off)
    abseege: to saw off
    aus: out (of)
    ausse: outside; vun ausse: from the outside
der Backeschtee: brick
 es Backeschteehaus, -heiser: brick house
der Backoffe, -effe: bake oven
der Bicherschank, -schenk: book case
 es Blettli, Blettin: saucer
die Bortsch, Bortsche: porch
    breet: broad, wide
 es Brennhols: firewood
 es Daageslicht: daylight

small table forks), and also spoons (tablespoons, tea
and coffee spoons, soup spoons, and ladles).

Up there my mother has her salt and pepper, sugar
and coffee, and her spices; and down there she has
pitchers, the coffee pot, and the tea kettle. The
large iron kettles she has in the summer-kitchen, and
that kitchen door leads out into the yard and to the
garden. (But don't walk through the screen door; one
has to open it first!)

Here is the sink. One turns here, and the water
comes out at this spigot. Then one shuts it off again
and washes the dishes in the washbasin with the dishcloth. One then dries them off. Here are also dish-
rags and towels.

The pantry is there behind that door. Our food-
stuffs are in there. And next to (beside) the kitchen
table stands my little brother's high chair.

Why is the kitchen table so large? I have seven
siblings!


```
 es Dachfenschder, -re:   dormer window
der Deckel:  cover, lid
     defor:  for it
     degeege:  against it
der Deller:  plate
     dick:  thick, fat
der Diergnobb, -gnebb:  doorknob
die Dierschwell:  door sill
der Dischlumbe:  dishrag
     dohin:  in here
die Drepp, Drebbe:  step, stair
     drin:  in there
     drinsitze:  to sit in (it)
     druffleige:  to lie on (it)
     druffschteh:  to stand on (it)
der Eckschank, -schenk:  corner cupboard
     eise (eisn + ending):  iron
```

die Esssache: foodstuff
die Essschtubb, -schtuwwe: dining room
    ewwerscht: uppermost, topmost
der Feierheerd: fireplace
der Feierhund: andiron
die Fens, Fense: fence
die Fenschderscheib, -scheiwe: windowpane
der Flohr: floor, story (of a building)
die Frantschtubb, -schtuwwe: front room
der Gaarde: garden
der Gang, Geng: hall, hallway
die Gawwel, Gaww(e)le: fork
    gell?: right? (see note)
 es Gewaerz: spices
 es Giwwelend, -enner: gable end
 es Glaas, Glesser: glass
der Glaasschank, -schenk: china closet, cabinet for
     glassware
der Graahne: spigot
    greessermache: to make larger
der Grundflohr: ground floor
 es Gscharr (Gschaerr): dishes
    gschwischich: between
der Haffe, Heffe: pot
 es Handduch, -dicher: towel
die Hausdier, -diere: house door
der Hausschlissel: house key
die Heemet, Heemede: home
    hinnenausgeh: to go out back
der Hinnerhof, -heef: back yard
    hinnich: behind, in back of
 es Hochschtiehlche, -cher: little high chair
der Hochschtuhl, -schtiehl: high chair
 es Hols: wood
die Holskischt: wood box
der Holsoffe, -effe: wood stove
der Kaerpet, Kaerpets (Karrepet): carpet
die Kaffikann, -kanne: coffeepot
der Kandell: spouting
die Katz, Katze: cat
der Kessel: kettle
die Kich, Kiche: kitchen
die Kichedier, -diere: kitchen door
der Kichedisch, -dische: kitchen table
die Kicheschank, -schenk: kitchen cupboard

die Kischt, Kischde: box, chest
der Kochoffe, -effe: cook stove
der Kohleoffe, -effe: coal stove
 es Koppli, Kopplin (Kopplicher): cup
der Laade, Leede: shutter
der Licktrickoffe, -effe: electric range
der Leffel, Leffle: spoon
     lengermache: to make longer
 es Lewesmiddel: foodstuff
die Lewesschtubb, -schtuwwe: living room
 es Maentelbord (Maentel): mantelpiece
die Meening, Meeninge: opinion
 es Messer, Mess(e)re: knife
die Mickedier, -diere: screen door
der Nascht, Nescht: branch
     nausfiehre: to lead out
     nausgeh: to go out
     neiduh: to put in
     neifiehre: to lead into
     neigeh: to go in(to)
     neischtecke: to stick in(to)
     nuffgeh: to go up
     nuffhupse: to jump up
 es Offerohr, -rohre: stovepipe
     paar: couple, few
die Pann, Panne: pan
der Parlor: parlor
der Peffer: pepper
 es Pitscher, Pitschers: pitcher
     rauskumme: to come out
     reibringe: to bring in
     reihole: to fetch in
     reikumme: to come in
     rumdrehe: to turn around
der Sals: salt
     sauwer: clean
der Schaddebaam, -beem: shade tree
der Schank, Schenk: cabinet, cupboard
der Schannschtee: chimney
die Schindel, Schindle: shingle
die Schkriendier, -diere: screen door
die Schissel, Schissle: bowl
der Schlissel, Schlissel: key
 es Schlisselloch, -lecher: keyhole
     schloofe: to sleep
der Schpeicher: second floor

die Schpielbank, -benk: sink
der Schpiellumbe: dish cloth
die Schpielschissel, -schissle: dishwashing bowl
der Schtock, Schteck: floor, story (of a building)
die Schtroos, Schtroosse: street
die Schubblaad, -laade: drawer
der Sessel: easy chair
 es Soofe: sofa
die Summerkich, -kiche: summer kitchen
    uffbasse: to look out, watch out
    uffmache: to open
    uffschliesse: to unlock
    ungefehr: about, approximately
    unne: without
    vedderscht: front-most
der Vorgang, -geng: foyer, front hall
der Vorhang, -heng: curtain
 es Wandbabier: wallpaper
    wedder: against
    weege: because of, on account of
    weeich: because of, on account of
die Wuhnschtubb, -schtuwwe: living room
der Wuffi: dog's name
    zidder: since
    zudrehe: to turn off
    zugucke: to observe, look at
    zumache: to close
    zwische: between

Vocabulary Notes

abhacke, abschpiele, abseege, nausfiehre, neischtecke,
  rumdrehe, reikumme, reihole, reibringe, rauskumme,
  uffmache, zudrehe, zugucke, zumache are conjugated
  like kumme.

abdrickle is conjugated like rechle.

drinsitze, nuffhupse, uffbasse, uffschliesse are conjugated like schwetze.

es Backeschteehaus: also es Backeschteenehaus, es
  Backeschteenichhaus.

die Bortsch: also die Portsch.

der Dischlumbe: also der Dischlappe.

gell? The PG speaker uses this particle after a state-
    ment (occasionally before) that he thinks his lis-
    tener will agree with. Die Schtubb iss schee, gell?
    "The room is nice, (isn't that) right?" The impli-
    cation is that the speaker believes the listener
    will say--or at least think--"Yes, I agree; it is
    nice."

der Graahne: also der Schpicket, -s.

die Mickedier: also die Schkriendier.

naeryeds: also narr(i)yeds.

die Schpielbank: also die Singk, der Sank; die Wesch-
    bank, der Wasserschank, die Wasserbank.

der Schtock: also der Flohr.

aerschde Schtock: also Grundflohr.

die Wuhnschtubb: also die Wohnschtubb, der Parlor, die
    Lewesschtubb, die Frantschtubb.

Grammar

A. Prepositions

1. We are already acquainted with many PG prepositions; every reading selection has included several.
Hopefully you heeded the advice that you simply accept
the accusative or dative case forms that follow these
prepositions. But even if you did, you were no doubt
curious as to why darrich was followed by der Schulhof
(darrich der Schulhof), and in was followed by em Schulhof (im Schulhof); or why in one sentence you read ins
Schulhaus and in another im Schulhaus. Let us now find
out the reasons for these variations.

2. PG prepositions fall into three groups: (a) those
that take objects in the accusative case, (b) those
that take objects in the dative case, and (c) those
that take objects in either the accusative or dative
cases, depending on the circumstances (but the circumstances are always definite and clear).

3. The following prepositions are <u>always</u> followed by objects in the accusative case:

bis	until (till), by, as far as
darrich	through
fer, for	for
geeich, geegge	against
um	around
unne	without
wedder	against

There are no exceptions to the fact that these prepositions take accusative objects, but the following comments concerning their use are necessary.

Physical contact is expressed by <u>wedder</u>, never by <u>geeich</u> or <u>geegge</u>:

Die Nescht waxe wedder es Haus.	The branches are growing against the house.
Ich hab nix geeich die Tietschern.	I have nothing against the teacher (f.).

<u>Bis</u> is often followed immediately by another preposition:

Die Fens geht bis zum (zu+em) Hinnerhof naus.	The fence goes as far as out to the back yard.

4. The following prepositions are <u>always</u> followed by objects in the dative case:

aus	out of	vun	from
bei	with, by, at	weeich or	about, because of,
mit	with	weegge	or on account of
noch	to, toward	zidder	since
nooch	after	zu	to

In the reading selection of this section you read the phrase "Meinre Mudder ihrer Meening nooch . . . ." The preposition <u>nooch</u> <u>follows</u> its object whenever it means "according to."

5. The following prepositions are followed by either accusative or dative objects:

an	at, to
hinnich or hinner	behind, in back of
in	in, into
iwwer	over, above

newich or newe	beside, next to, near
owwich	over, above
uff	on, upon
unnich	beneath, under, among
ver or vor	before, in front of

6. The objects of these prepositions must appear in the accusative case if action <u>toward</u> or <u>into</u> the area represented by the object is implied. For instance, one would not say, <u>Ich geh ins Haus</u> if one were already <u>in</u> the house. The implication is that one starts going into the house while still outside it, thus starts toward the house and then moves <u>into</u> the area called "house."

Thus as you move into the area called <u>Schulhof</u>, you say, <u>Ich geh in der Schulhof</u>. As you move into the area called <u>Schulhaus</u>, you say, <u>Ich geh ins (in+es) Schulhaus</u>. As you move into the area called <u>Schulschtubb</u>, you say, <u>Ich geh in die Schulschtubb</u>.

The verb, then, is important in determining whether one uses the accusative or dative cases after these particular prepositions. Indeed, some verbs strongly suggest movement from one area into another; <u>geh</u> "to go," <u>renne</u> "to run," <u>faahre</u> "to drive," and <u>laafe</u> "to walk," are just a few of such verbs. Chances are very good that one of these "two-way" prepositions used in conjunction with such a verb must then be followed by an object in the accusative case.

7. However, a verb that indicates motion does not necessarily imply motion toward or into; it may simply express motion <u>within</u> an area. Motion within an area, or no motion at all, both dictate the use of a dative object. If children are running (around) in the school yard, we say, <u>Die Kinner renne im Schulhof rum</u>. The separable prefix <u>rum</u> is the equivalent of E "around," and is used almost always if the action takes place within an area:

| Er laaft in der Schtubb rum. | He is walking around in the room. |
| Er rennt im Haus rum. | He is running around in the house. |

The verbs of these two clauses are not <u>laafe</u> and <u>renne</u>, but <u>rumlaafe</u> and <u>rumrenne</u>.

8. If a verb expresses no motion, then these two-way prepositions are followed by dative case objects:

Er schteht in der   He stands in the room.
Schtubb.

Sitze "to sit," hocke "to sit," leigge "to lie," are other examples of verbs that express no motion.

9. Now let us examine the differences between, and the reasons for, such forms as geegge, degeegge; mit, demit; and iwwer, driwwer. The complete list of these forms follows:

Accusative
   darrich, dedarrich   um, drum
   for, defor   wedder, dewedder
   geegge, degeegge

Dative
   aus, draus   vun, devun
   bei, debei   weegge, deweegge
   mit, demit   zu, dezu
   nooch, denooch

Accusative or Dative
   an, draa   uff, druff
   in, drin   unnich, drunner
   iwwer, driwwer   vor, devor
   newe, denewe   zwische, dezwische

Note that dedarrich is the only form ending in -ich. Also note that de- precedes prepositions beginning with consonants, and dr- precedes prepositions beginning with vowels.

The prepositions without the prefixes are used:

(a) Whenever the object is a noun.

Er schteht uff em   He stands on the roof.
Dach.
Er geht mit der Memm. He goes with his mother.

(b) Whenever the object is a pronoun whose antecedent is a living thing.

Er geht mit mir   He goes into town with me.
(dihr,ihm,ihre,uns,
eich,ihne) ins
Schteddel.

Er hot nix geegge mich   He has nothing against
(dich,ihn,sie,uns,eich,   me.
sie).

10. Prepositions with the prefixes de- and dr- must
be used whenever their objects are pronouns whose ante-
cedents are inanimate. In such instances, the prefix
takes the place of the pronoun, and the pronoun is
dropped entirely.

For instance, we say Die Nescht waxe wedder es Haus
"The branches grow against the house"; but Sie waxe dewedder "They grow against it." Er schteht uff em Dach
"He stands on the roof"; but Er schteht druff "He
stands on it." Sitzt die Gremmemm im Schockelschtuhl?
"Is Grandma sitting in the rocker?"; but Ya, sie sitzt
drin "Yes, she's sitting in it."

11. Uff em Offe druff "On the stove (thereon)," and
im Backoffe drin "in the oven (therein)," and all other
similar expressions are not considered redundancies by
the PG speaker. These are perfectly acceptable expressions.

B. Inseparable Prefixes

We have been reading and using inseparable prefixes
since Part One of Lesson One. A vocabulary note informed us that be-, er-, ge-, and ver- were four common
prefixes that never separate from the verb and are
never accented. To be sure, there are other words that
can be used as inseparable prefixes, but they are not
always inseparable; verbs with such prefixes are best
learned as they come up in the vocabularies.

We also learned that an inseparable prefix changes
the meaning of a verb but not the conjugation. Here is
an example:

schteh "to stand"   verschteh "to understand"
   ich schteh   ich verschteh
   du schtehscht   du verschtehscht
   er schteht   er verschteht

Thus if you know the conjugation of a verb without
an inseparable prefix, you will know its conjugation
with an inseparable prefix, but it is best to treat the

two verbs as entirely different and separate vocabulary words. Only then can you be certain to learn the correct E equivalent for both.

The same holds true for verbs with and without separable prefixes.

C. Separable Prefixes

1. In English, we can understand something, but we can also stand under something; we can overlook something, but we can also look over something; we can upset something, but we can also set up something. If we equate understand, overlook, and upset with PG verbs beginning with inseparable prefixes, then we can equate stand under, look over, and set up with PG verbs beginning with separable prefixes. Look over is especially like PG verbs with separable prefixes. We can say, for example, he looked the car over and then bought it. Note that <u>over</u> follows the object <u>car</u>; and that is exactly what happens with separable prefixes in PG main clauses whose verbs are in the simple present tense (and the simple past tense form <u>waar</u>). Not only are separable prefixes always accented but they actually separate from the verb and go to the very end of the sentence (clause). The reading selections for this lesson contain several verbs with separable prefixes (the verbs in their infinitive forms then follow in parentheses). Some examples follow:

Mer gehne ins Haus nei.	We're going into the house.
Mer mache die Dier uff.	We open the door.
Es Wasser kummt am Graahne raus.	The water comes out at the spigot.

2. If you remember what to do with "double" verbs such as <u>yaage geh</u>, <u>schiesse heere</u>, or <u>geh losse</u> (Chapters 4 and 5) then you already have a good idea what to do with separable prefixes. For example, let's compare <u>yaage geh</u> and <u>neigeh</u>:

Er geht im Harrebscht yaage.
Er geht ins Haus nei.

Er heert die Schitz schiesse.
Er macht die Dier uff.

Er losst sei Keit in die Heeh geh.
Er geht   die   Dreppe   nuff.

And if you remember how these verbs were treated
when they were dependent upon a modal auxiliary, you
are "home free."

Eldre solle ihre Kinner schpiele losse.
Er soll   ins   Haus   neigeh.

The idea, you see, is exactly the same; the only difference lies in the fact that under no circumstances do
separable prefixes separate from their verbs in their
infinitive forms, the form of a verb dependent upon a
modal auxiliary.

The various prepositions we studied in the beginning
of this grammar section are very often used as separable prefixes, but PG also uses adverbs, adjectives,
nouns, and even verbs.

3. You no doubt noticed that some separable prefixes
appear in "couplets"; for example, nei,rei; nuff,ruff;
nunner,runner. The forms beginning with n express
movement away from the speaker or the speaker's area.
The forms beginning with r express movement toward the
speaker or speaker's area.

Let's imagine two people, A and B. Let's further
imagine that A is inside the house and B is outside.
If B enters the house, he thinks to himself, Ich geh
ins Haus nei "I'm going into the house." B uses nei
because, as far as he is concerned, he is moving away
from the area outside the house. A thinks to himself,
B kummt ins Haus rei "B is coming into the house." A
uses rei because, as far as he is concerned, B is moving toward and into A's area inside the house.

Now imagine that A is standing at the bottom of the
stairs and B is standing at the top. If A starts up
the stairs, he thinks to himself, Ich geh die Schteeg
nuff "I'm going up the stairs." A uses nuff because he
is leaving his original area at the bottom. B thinks
to himself, A kummt die Schteeg ruff "A is coming up
the stairs." B uses ruff because A is coming toward
him.

Conversely, if B came down the stairs towards A, A

would say, <u>Er kummt die Schteeg runner</u> "He's coming
down the stairs"; and B would say, <u>Ich geh die Schteeg
nunner</u> "I'm going down the stairs."

By the way, <u>Er geht ins Haus nei</u> "He goes into the
house (in)," or <u>Er kummt aus em Haus raus</u> (and all
other similar expressions) merely <u>sound</u> redundant to
the E speaker. They are all perfectly good expressions
in PG.

4. <u>hie</u> and <u>haer</u>

If we have understood the preceding explanation of
<u>rei</u> and <u>nei</u> above, we should have no difficulty with
<u>hie</u> and <u>haer</u>, two particles that are very often them-
selves used as separable prefixes. <u>Hie</u> (Standard German <u>hin</u>) always indicates motion away from the speaker;
<u>haer</u> (Standard German <u>her</u>) always indicates motion
toward the speaker:

Er geht ans Haus hie.   He goes to the house.
Er kummt vum Haus haer.   He comes from the house.

The first sentence indicates that the actor leaves
either his original position or the speaker's position
and moves away toward the house. The second sentence
indicates that the actor leaves his original position
at the house and moves toward the speaker.

5. The <u>n</u> of <u>nei</u>, <u>nunner</u>, etc., is the <u>n</u> of St.G. <u>hin</u>
which has disappeared from PG <u>hie</u>. The <u>r</u> of <u>rei</u>, <u>run-
ner</u>, etc., is the final <u>r</u> of St.G. <u>her</u>, PG <u>haer</u>.

StG   hinein   PG nei
      hinunter   nunner

StG   herein   PG rei
      herunter   runner

CHAPTER SIX

Part Two

Well, now we know something about the first floor of our young friend's house and the grammar needed to express these ideas. Let's continue with our guided tour by climbing the stairs to the second floor, the Schpeicher.

Oft wann mer weit iss vun deheem
Kumme Gedanke iwwer eem
Wie mer's deheem so gut hot ghatt
Un wie die Mammi oft eem gsaat:
*   "Waart--draus iss net deheem!"*

Mir Buwe hen als yuscht gelacht
Un unser Gschpichde fattgemacht;
Vun heem geh hen mer net geahnt,
Nau bin ich's awwer gut bekannt
*   Dass draus iss net deheem.*

Bleib, bleib deheem, O Kind! Es Nescht
Ver's Veggeli iss doch's allerbescht;
Draus in der raue, weite Welt
Sin Schtarrem un Blitz un groossi Kelt--
*   Ach, draus iss net deheem!*[1]

---

[1]*Charles C. Ziegler, from "Drauss un Deheem,"* Drauss un Deheem.

Es Haus; Es Zwett Deel

Kumm, mer gehne nau widder in der Gang naus, un die
Schteeg nuff zum Schpeicher (zwedde Schtock). Datt
drowwe sin die Schloofschtubbe (Bettschtubbe). Ya, die
alt Schpeicherschteeg hot viel Drebbe; die Decke sin
hoch in so me alde Bau(e)rehaus wie des. Es G(e)lenn-
der helft eem.

Endlich. Do howwe iss aa die Baad(e)schtubb. Waart
en Minutt, ich dreh 's Licht aa. Nau kenne mer besser
sehne. Du kannscht rumgucke, awwer du sehnscht nix
Neies. In eenre Eck vun der Baadeschtubb sehnscht en
Weschbohl, un in der anner Eck devun iss en Toilet.
Datt iss nadierlich der Baadezuwwer; mer nemmt sei Baad
do drin. Un do sin die Seef, Weschlumbe, Handdicher.
Der Schpiggel iwwer der Weschbohl iss die Dier fer der
Medizienschank.

Un do sin die Bettschtubbe. Die vedderscht Schtubb
iss meine Eldre ihre Schloofschtubb, am End vum Gang.
Guck mol nei. Die Schtubb iss viel greesser ass meini.
Un des Bett iss es greesscht im Haus.

Sehnscht der Gleederschank? Schee, gell? Un der
Bierro iss aa net glee. Datt in der Eck iss meinre
Schweschder ihre Kinnerwieg. Oweds schtellt mei Memm
die Wieg newwich es Bett, un marriyeds schtellt sie es
widder zrick in die Eck.

Die Kischt mit Veggel un Dullebuhne druff iss fer
der Mudder ihre Debbiche. Sie iss arrig alt. In der
Gleederkammer sin net yuscht Gleeder drin, awwer aa
Leindicher (Bettdicher), Koppekisse, Bettdecke, un en
Fedderbett.

Was? Ya, du hoscht recht. Die Memm iss immer am
Abschtaawe; alles iss schee sauwer. In dem Haus iss
nie ken Schtaab. Die anner vier Bettschtubbe sin fer
uns Kinner. Die do iss fer mei zwee greessere Brieder
un mich. Ich schloof datt in dem Drunnelbett; die Maddratz iss wennich dinn.

Wu geht die Dier hie? Hinner daerre Dier iss die
Schteeg zum ewwerschde Schpeicher. Mer hen yuscht alt
Hausrot (alde Haussache) do drowwe. Mer gehne net oft
nuff. Im Summer iss es zu heess, un im Winder zu kalt.

Nau gehne mer nunner zum Keller. . . . Do simmer

The House; Part Two

Come, we'll go now out into the hall again and up
the stairs to the second floor. Up there are the bedrooms. Yes, the old stairway has many steps; the ceil-
ings are high in such an old farmhouse like this. The
railing helps one (helfe + dative).

Finally. Up here is the bathroom. Wait a minute,
I'll turn the lights on. Now we can see better. You
can look around, but you won't find anything new. In
one corner of the bathroom you see the sink (washbowl),
and in the other corner of it is a toilet. There is
naturally the bathtub; one takes his bath in it. And
here are the soap, washcloth, towels. The mirror over
the washbowl is the door for the medicine cabinet.

And here are the bedrooms. The front-most room is
my parents' bedroom at the end of the hall. Look in
(once). The room is much larger than mine. And the
bed is the biggest in the house.

Do you see the clothes cabinet? Nice, huh? And the
bureau is also not small. There in the corner is my
sister's cradle. In the evenings my mother places the
cradle next to the bed, and in the mornings she places
it back into the corner.

The chest with the birds and tulips on it is for my
mother's quilts. It is very old. In the clothes
closet are not only clothes, but also linen sheets
(bedsheets), pillows, bed covers, and a feather bed.

What? Yes, you're right. Mom is always dusting;
everything is always nice and clean. In this house
there is never any dust. The other four bedrooms are
for us children. This one is for my two older (bigger)
brothers and me. I sleep there in the trundle bed; the
mattress is a little thin.

Where does this door go? Behind that door is the
stairway to the attic. We have only old furniture (old
house furnishings) up there. We don't go up there
often. In summer it is too hot, and in winter too cold.

Now we'll go down to the cellar. . . . Here we are

widder hunne im aerschde Schtock, un do iss die Kellerdier. Die Kellerschteeg iss net so hoch ass die
Schteeg zum Schpeicher, awwer geb acht, sie iss enger
ass wie die Schpeicherschteeg.

Des iss die nei Fanness. Sie brennt Eel; mer
brauche die Esch nimmi nausdraage. Datt sin der Memm
ihre Weschmaschien un Weschzuwwer. Die Weschschpelle
sin in dem do gleene Karreb. Die Weschlein iss draus
im Hof.

Datt iss es Kellerloch. Mer kenne die paar Drebbe
nuff, der Kellerschlaag uffmache, un in der Hinnerhof
nausgeh.

Kumm doch marriye widder verbei. Dann weis ich der
die Gebeier hinnedraus im Hof.

Vocabulary

abschtaawe: to dust off
achtgewwe: to watch out, be careful
es Baad: bath
die Baad(e)schtubb, -schtubbe: bathroom
der Baadezuwwer, -ziwwer: bathtub
es Bauerehaus, -heiser: farmhouse
die Bettdeck, -decke: bed cover
es Bettduch, -dicher: bed sheet
die Bettschtubb, -schtuwwe: bedroom
der Bierro, Bierros: bureau
der Debbich, Debbiche: quilt, bedspread
dinn: thin
es Drunnelbett, -bedder: trundle bed
die Dullebuhn, Dullebuhne: tulip
es Eel: oil
es End: end
eng: narrow
die Esch: ashes
die Fanness: furnace
die Fedderdeck, -decke: feather bed
es Gebei, Gebeier: building
es G(e)lennder: bannister
die Gleederkammer: closet
der Gleederschank: clothes cabinet, wardrobe
die Haussache (pl.): house furnishings
es Hausrot: furniture

again down on the first floor, and here is the cellar
door. The cellar stairs are not as high as the stairs
to the second floor, but watch out, it is narrower than
the stairs to the second floor.

This is the new furnace. It burns oil; we don't
have to carry out the ashes. There are Mom's washing
machine and wash tub. The clothespins are in this
little basket. The washline is out in the yard.

There is the cellar doorway. We can (go) up a few
steps, open the outside cellar door, and go out into
the back yard.

Come by again tomorrow. Then I'll show you the
buildings out back in the yard.

hinnedraus: out in back
howwe: up; do howwe: up here
hunne: down (here)
der Keller: cellar
die Kellerdier, -diere: cellar door
 es Kellerloch, -lecher: cellar doorway leading outside
der Kellerschlaag: outside cellar door
die Kellerschteeg, -e: cellar stairs
die Kinnerwieg: cradle
die Kischt, Kischde: chest
 es Kopp(e)kisse: pillow
 es Leinduch, -dicher: bedsheet
die Maddratz, Maddratze: mattress
der Medizienschank, -schenk: medicine chest, cabinet
    nausdraage: to carry out
    nunnergeh: to go down
    recht: right
    recht hawwe: to be correct
    rumgucke: to look around
die Schloofschtubb, -schtuwwe: bedroom
der Schpiggel, -e: mirror
der Schtaab: dust
die Schteeg, -e: stairs
die Seef: soap
die Toilet, Toilets: toilet
    verbeikumme: to come by, "drop" by
der Voggel, Veggel: bird

```
        weise:  to show
die Weschbohl:  washbowl
die Weschlein:  washline
der Weschlumbe:  washcloth
die Weschmaschien, -e:  washing machine
die Weschschpell, -schpelle:  clothespin
der Weschzuwwer, -ziwwer:  washtub
die Wieg:  cradle
```

Vocabulary Notes

rumgucke is conjugated like kumme.

abschtaawe, achtgewwe are conjugated like schreiwe.

weise is conjugated like schwetze.

der Bierro: also der Buro, -s. Der Aarichtdisch is the dresser.

es Drunnelbett: also es Bettleedel, es Gleebett.

die Dullebuhn: also Dullebaan, Dollebaan, Dulleblumm.

die Weschschpell: also die Weschglamm.

die Wieg: also die Schockel.

CHAPTER SEVEN

Past Participles
and the
Present Perfect Tense

With but one exception, all verbs in our first six
lessons were in the present tense. Let us now see how
the PG speaker expresses himself in the past tense. In
a sense, this is also a review lesson. We will be re-
studying the reading selections of Chapters One through
Six and changing the present tense verbs to the present
perfect. We thus have a chance to review and add to
our store of vocabulary words.

*Der Tscheeck Desch iss gschtarrewe; der
Tscheeck waar aa alt genunk fer schtarrewe;
sexunachtsich Yaahr iss en scheeni Elt. 'S
waar net, dass die Leit em Tscheeck's Lewe
absaage hen wolle, dass sie net waarde hen
kenne uff sei Dot, wie er mol fascht im Bett
gelegge hot mit ken abbaddichi Granket ass
wie Aldersschweche. Awwer der Wunnerfitz hot
sie gebloogt, was des fer en Leicht gebt,
wann er mol schtarrewe sett. Der Tscheeck
hot nix liewer geduh, ass wie an en Leicht
geh, un verschtanne, aa zu bleiwe fer's Esse.
Un was hot er esse kenne!*[1]

---

[1]*Lloyd A. Moll, from "Em Tscheek Desch Sei Leicht,"
Am Schwarze Baer.*

145

Past Participles and the Present Perfect Tense

We have seen that our knowledge of the present tense enables us to express the future tense, too, especially if we use an adverbial expression of future time, e.g., Marriye gehne mer ins Schteddel "Tomorrow we will go (are going to go) into town." But gaining a working knowledge of the past tense will not be quite so easy. The PG speaker expresses action in the past by using the present perfect tense, a combination of an auxiliary verb (either hawwe or sei) and the past participle of the main verb.

A. The Formation of Past Participles

For the purpose of forming past participles, verbs are divided into two groups, called weak and strong. The weak verbs approximate E regular verbs (live,lived; farm,farmed) and the strong verbs approximate E irregular verbs (drive,driven; rise,risen). Much like E, PG weak past participles end in -t, and strong past participles end in -e (as usual, the n has disappeared). But there is more to a PG past participle than just the endings.

1. Let us first of all consider the weak verbs, which, in turn, have weak past participial forms. Generally speaking, we drop the infinitive ending -e from the verb, place ge- (or g-) in front of the resulting stem, and attach the -t ending:

$$\text{ge + stem + t}$$

mache   ge + mach + t = gemacht
kenne   ge + kenn + t = gekennt

Weak verbs ending in a consonant + le (e.g., rechle) drop the final e and add an e before the l, then add the ending t: rechle,gerechelt; weissle,geweisselt.

Weak verbs whose stems end in d (e.g., deide) merely change the d to t: deide,gedeit.

Weak verbs whose stems end in dd (e.g., andwadde) merely change the dd to tt: andwadde,geandwatt.

2. It follows, then, that PG strong verbs are those that have the past participial ending -e. Generally

speaking, the participle of a strong verb is a combination of <u>ge</u> + stem + <u>e</u>: <u>lese,gelese</u>; <u>backe,gebacke</u>. But many strong verbs change their stem vowel, as the following verbs and their past participles illustrate: <u>ringe,gerunge</u> (<u>i</u> to <u>u</u>); <u>bleiwe,gebliwwe</u> (<u>ei</u> to <u>i</u>); <u>nemme,genumme</u> (<u>e</u> to <u>u</u>).

Knowledge of such stem vowel changes is absolutely necessary if one is to construct a correct past participle of a strong verb. There is but one answer: the past participle of strong verbs must be committed to memory.

3. Whether a verb be strong or weak, we must note these exceptions to our general rules for the formation of past participles.

(a) The B/B system of spelling adds the prefix <u>g</u>- (no <u>e</u>) to stems beginning with the letters <u>h</u>, <u>f</u>, and <u>s</u> (<u>sch</u>): <u>heesse,gheesse</u>; <u>fuule,gfuult</u>; <u>sitze,gsotze</u> (<u>gsesse</u>); <u>schwetze,gschwetzt</u>. The combination <u>gh</u>- is pronounced like <u>k</u>.

(b) The prefix <u>ge</u>- is dropped altogether before some stems beginning with <u>g</u>: <u>gewwe,gewwe</u>; <u>geh,gange</u>; <u>griegge,grickt</u>. But it would be best to think of these as exceptions and not the general rule.

Others (e.g., <u>kumme</u>, <u>warre</u>) are best learned as they come up in the vocabularies.

(c) A verb ending in -<u>iere</u> does not usually add the prefix <u>ge</u>- in PG literature: <u>bassiere,bassiert</u>; <u>buchschtaawiere,buchschtaawiert</u>. But many PG speakers use the prefix <u>ge</u>- even before the past participles of such verbs.

(d) No <u>ge</u>- is added to a verb with an inseparable prefix: <u>verschteh,verschtanne</u>; <u>bezaahle,bezaahlt</u>. (Often the <u>e</u> of the prefix is dropped before <u>h</u>, <u>s</u>, and <u>sch</u>: <u>bhalde</u>, <u>bsuche</u>. <u>bh</u>- is then pronounced like <u>p</u>.)

(e) If a verb has a separable prefix, the prefix is simply attached to the front of the past participle; the entire construction is then written as one word: <u>uffschteh,uffgschtanne</u>; <u>rumrenne,rumgerennt</u>.

B. The Use of the Past Participle
in the Present Perfect Tense

1. We have noted that waar is the only simple past
tense verb form in PG. In every other instance, the
past is rendered by means of the present perfect tense,
a compound tense made up of a past participle and its
auxiliary, hawwe or sei.

2. The auxiliary is normally the second element of a
clause. Thus if a sentence (clause) consists of nothing more than a subject and verb, the past participle
immediately follows the auxiliary: Die Schulbell ringt;
die Schulbell hot gerunge.

If a sentence or clause consists of a subject and a
predicate containing more than just a verb, then the
auxiliary again takes second position and the past participle goes to the very end: Die Schulkinner gehne
ins Schulhaus, die Schulkinner sin ins Schulhaus gange;
die Schulschtubb hot vier Wend, die Schulschtubb hot
vier Wend ghatt (ghadde).

3. If a modal has no dependent verb, then it is
treated like any other verb; its past participle goes
to the end: Er kann Deitsch, er hot Deitsch gekennt.

But if a modal auxiliary has a dependent verb, then
both modal and dependent verb go to the end in their
infinitive forms, first the dependent verb and then the
modal: Er kann Deitsch schwetze, er hot Deitsch
schwetze kenne; Die Kinner misse in die Schul neigeh,
die Kinner hen in die Schul neigeh misse.

4. Now we need only choose between the auxiliaries
sei and hawwe, and we can begin converting the first
reading selection of Chapter One, Es Schulhaus.

(a) The auxiliary sei is used with verbs that have
no direct objects (intransitive verbs) and show a
change of position or condition. Both conditions must
be met, not just one or the other. Er geht ins Haus;
er iss ins Haus gange. Sie renne im Schulhof rum; sie
sin im Schulhof rumgerennt.

(b) The auxiliary hawwe is used with verbs that have

direct objects (transitive verbs), and also with intransitive verbs that do not show change of position or condition: <u>Der Meeschder kennt die Kinner</u>; <u>der Meeschder hot die Kinner gekennt</u>. <u>Die groosse Kinner hocke links; die groosse Kinner hen links ghockt</u>.

5. <u>Sei</u> is always the auxiliary of <u>sei</u>: <u>Er iss g(e)west</u>.

<u>Hawwe</u> is always the auxiliary of <u>hawwe</u>: <u>Er hot Bicher ghatt (ghadde)</u>.

<u>Hawwe</u> is the auxiliary of all modals: <u>Er hot in die Schul gemisst</u>.

6. Nothing we have said so far about the present perfect tense changes in a sentence or clause employing inverted word order. The auxiliary still maintains its second position in the sentence and the past participle still goes to the end.

Marriye gehne mer ins Schteddel.
Geschder sin mer ins Schteddel gange.

Marriye misse mer ins Schteddel.
Geschder hen mer ins Schteddel gemisst.

Marriye misse mer ins Schteddel neifaahre.
Geschder hen mer ins Schteddel neifaahre misse.

And if it should happen that a double infinitive is dependent on a modal auxiliary, we end up with three infinitives in the present perfect:

Ich kann denowed danse geh.
Ich hab denowed danse geh kenne.

7. Now let us see how the first reading selection of Chapter One "translates" into the present perfect tense. Remember: the PG present perfect is the equivalent of the E simple past tense.

These are the past participles of verbs used in the reading.

Weak:		Strong:	
frooge	gfroogt	geh	gange
hawwe	ghatt,	heesse	gheesse
	ghadde	henke	ghanke, ghunke
hocke	ghockt	(weak if transitive)	

Weak:		Strong:	
kenne	gekennt	kumme	kumme
kenne(mod.)	gekennt	lese	gelese
lanne	gelannt	ringe	gerunge
saage	gsaat	schreiwe	gschriwwe
schwetze	gschwetzt	schteh	gschtanne
sei	g(e)west	sitze	gsotze, gsesse
wisse	gewisst	verschteh	verschtanne
wuhne	gewuhnt	(inseparable prefix--	
yuuse	geyuust	no ge-)	

The predicate of Die Schulbell ringt contains only
the verb; the past participle follows right after the
auxiliary: Die Schulbell hot gerunge. Hawwe is the
auxiliary even though ringe is intransitive, because
no motion is implied.

The predicate of Die Schulkinner gehne ins Schulhaus
contains more than just the verb; the past participle
goes to the end: Die Schulkinner sin ins Schulhaus
gange. Sei is the auxiliary because geh is intransitive and expresses motion.

Es Schulhaus hot yuscht ee Schtubb ghatt (ghadde).
The auxiliary of hawwe is hawwe.

Die groosse Kinner hen rechts ghockt. The auxiliary
is hen even though hocke is intransitive, because no
motion is expressed. Thus sitze also takes hawwe as
its auxiliary.

Er schteht uff un saagt, "Gude Marriye . . ." is a
bit more involved. Uff is a separable prefix; the verb
is uffschteh, an intransitive verb implying motion.
But saage is transitive (whatever is said is the direct
object of "say") and we must thus use hawwe as its auxiliary. Therefore, we cannot use one auxiliary for
both verbs: Er iss uffgschtanne un hot gsaat, "Gude
Marriye, Klass."

"Heit iss der aerscht Schuldaag g(e)west." The verb
sei always takes the auxiliary sei.

"Ich kann Englisch schwetze," contains a modal and
dependent verb: "Ich hab Englisch schwetze kenne." The
modal and dependent verb go to the end as infinitives.
All modals take hawwe as their auxiliary.

The strong verbs used in Part Two of Chapter One and in Chapters Two through Six follow:

Chapter 1, Part 2
bhalde, bhalde
bleiwe, gebliwwe
duh, geduh
gleiche, gegliche
hewe, ghowe
laafe, geloffe
nemme, genumme

Chapter 2, Part 1
hawwe, ghatt (ghadde)
warre, warre

Chapter 2, Part 2
backe, gebacke
gewwe, gewwe
gleiche, gegliche
helfe, gholfe
leigge, gelegge
schpringe, gschprunge
sehne, gsehne
ziehge, gezogge

Chapter 3, Part 1
no new strong verbs

Chapter 3, Part 2
nemme, genumme

Chapter 4, Part 1
aaduh, aageduh
faahre, gfaahre
gewwe, gewwe
heemkumme, heemkumme
uffschteh, uffgschtanne
wesche, gewesche

Chapter 4, Part 2
drinke, gedrunke
ringe, gerunge
rufe, gerufe

Chapter 5, Part 1
esse, gesse
fliege, gflogge
losse, gelosse
schiesse, gschosse
schwimme, gschwumme
waere, gewore

Chapter 5, Part 2
falle, gfalle
friere, gfrore
rumschpringe, rumgschprunge
schmeisse, gschmisse
schmelse, gschmolse
waxe, gewaxe

Chapter 6, Part 1
schliesse, gschlosse
schloofe, gschloofe
sehne, gsehne

Chapter 6, Part 2
draage, gedraage
weise, gewisse

Modals
darrefe, gedarreft; daufe
    gedauft; daerfe, gedaerft
kenne, gekennt
meegge, gemeecht
misse, gemisst
solle, gsollt, gsott;
    selle, gsellt, gsett
wolle, gewollt

PART TWO

CHAPTER EIGHT

Part One

The farmhouse and the barn each have certain handi-
ly accessible outbuildings grouped about them. There
are no hard and fast rules concerning these buildings;
some farms no doubt have all the buildings mentioned
in this lesson, while others may have no more than a
house, a barn, and possibly one or two others. Also,
there are no hard and fast rules concerning their
placement in one area or another. The following account of the buildings on our young guide's farmstead
is thus well within the realm of possibility.

O! wann ich yuscht en Bauer waer,
Un hett en gut Schtick Land,
Dann hett ich aa mei Seck voll Geld,
Un aa noch in der Hand.

Im Friehyaahr seet er's Hawwerfeld,
Un blanst, aafangs im Moi,
Grummbeere, Welschkann un so Sach;
Im Yuli macht er Hoi.

Der Bauer schafft im Summer hatt,
In Hoiet, Aern un Seehe,
Un bloogt sich arrick in der Sunn,
Muss schwitze viel im Meehe.

Doch kann er in sei Gaarde geh
Un brauch sei Sach net messe,
Un alle Daag frischer Selaat
Un gude Arrebse esse.[1]

[1]*David B. Brunner, from "Wann ich Yuscht en Bauer*
Waer," PG Verse.

155

Der Hof

Am neegschde Daag iss mei Freind, der Tschan, widder verbeikumme, un mer sin glei in die Summerkich neigange. Datt schteht der grooss Kicheoffe aus schwazem
Eise. Er hot noch nie so en groosser, alder, eisner
Kochoffe gsehne ghatt (ghadde). Der Offe hot mol in
der Kich gschtanne ghatt, un dann hen mei Eldre die
Summerkich graad hinne an der Kich aabaue losse, un
hen daer alt Holsoffe in die Summerkich neigschtellt.

Die Gremmemm kocht immernoch uff dem Offe, abbaddich
im Summer; dann watt's net so heess in der Kich. Dann
un wann darrefe mir Kinner aa datt drin esse.

Mer gehne naus zum Hinnerhof. Net weit weck vun der
Summerkich, zwische em Groossdaadihaus, wu mei Grooss-
eldre wuhne, un em Weschheisel, iss der Gaarde. Die
Gaardefens iss schee weiss; friehyaahrs bin ich immer
am Weissle, net yuscht die Gaardefens, awwer aa die
Fenseriggel un Poschde vor em Haus.

Es Weschheisel iss graad en glee(nes) Gebei. Mei
Memm hot schunn lang nimmi drin gewesche; du hoscht yo
die Weschmaschien im Keller gsehne. Es iss nau en
Schubb fer Reche, Schaufle, Schibbe, Schubbkarrich, un
Graasmaschien. Mei Groosseldre brauche so Sache, wann
sie gaerdle.

Do schteht aa en alder Schleifschtee drin. Der
Gremmpaepp hot als sei Sense un Sichle demit geschlif-
fe. Der Paepp schleift aa sei Ax un Beil do druff.

Un datt driwwe waar mol es Schpringhaus. Mei Gremmpaepp hot es graad iwwer en Schpring gebaut ghatt, un
es waar aa sei Grundkeller, un die Gremmemm hot als
ihre Millich do neigeduh. Des waar awwer lang zrick.

Dann hot der Gremmpaepp en Brunne gegraawe un hot es
Schpringhaus weckgenumme. Schpeeder hot er datt 's
Bumphaus gebaut. Nau kummt's Wasser vun gans weit
drunne im Grund. Die Bump muss es ruffbumbe un ins
Haus neibumbe.

Un des iss es Holsheisel. Mir hen als viel Hols ge-
braucht, un es Heisel waar immer voll. Mei Paepp hot's
Hols in en Holskarreb neigeduh un dann in die Kich nei-
gedraage. Die Holskischt hinnich em Offe hot immer
voll sei misse. Awwer nau brenne mer net viel Hols,

The Farmyard

On the next day my friend John came by again, and
we went right into the summer kitchen. There stands
the large kitchen stove of black iron. He had never
seen such a large, old iron cookstove. The stove had
once stood in the kitchen, and then my parents had the
summer kitchen built right in back at the kitchen, and
placed that old wood stove into the summer kitchen.

Grandma still cooks on that stove, especially in
summer; then it doesn't get so hot in the kitchen. Now
and then we children may also eat in there.

We go out to the back yard. Not far away from the
summer kitchen, between the house where my grandparents
live and the wash house is the garden. The garden
fence is nice and white; in the spring I am always
whitewashing, not just the garden fence, but also the
fence rails and posts in front of the house.

The wash house is just a small building. My mom
hasn't washed in it for a long time; you saw the washing machine in the cellar. It is now a shed for rakes,
shovels, spades, wheelbarrow, and grass mower (lawnmower). My grandparents need such things when they
garden.

In there stands also an old grindstone. Grandpop
used to grind his scythes and sickles with it. Dad
also grinds (sharpens) his ax and hatchet on it.

And over there was once the spring house. My granddad had built it right over a spring, and it was also
his ground cellar, and grandma used to put her milk
into it. But that was a long time ago.

Then granddad dug a well and took away the spring
house. Later he built the pumphouse there. Now the
water comes from far down in the ground. The pump
must pump it up and pump it into the house.

And this is the woodshed. We used to use a lot of
wood, and the shed was always full. My pop put the
wood into a wood basket and carried it into the kitchen. The woodbox behind the stove had to be full all
the time. But now we don't burn much wood, and the

un 's Heisel iss fascht leer. Mer brauche 's Hols fer
der Feierheerd in der Wuhnschtubb un fer der Gremmemm
ihre Holsoffe. Die Seeg, der Seegbock, un der Hackglotz sin immernoch drin.

Awwer mer hen aa viel Hickerihols fer sell Schmookhaus datt driwwe; es iss graad am Schlachthaus aagebaut.
Mei Gremmpaepp hot immer selwer(t) gschlacht; sei Vadder hot's Schlachthaus gebaut ghatt. Noh hot der Paepp
zwee gude Tietscher ghatt.

Er yuust noch all die alde Schaawer un der alt
Briehdroog un Waschder; aer un die Memm mache Lewwer-
wascht un Brotwascht graad wie Yaahre zrick. Un er
schmookt es Schunkefleesch yuscht mit Hickerihols. Uff
daerre do Bauerei hen mer der bescht Schpeck un Summer-
wascht in der Kaundi. Die Biefschteecks sin aa net
schlecht, kann ich der saage.

Nau hawwich awwer Hunger. Kumm, mer gehne zrick ins
Haus. Verleicht gebt uns die Memm heemgemachter Pannhaas odder Zidderli fer Middaag.

Des Heisel dat driwwe? Ya, des weescht net? Des
iss unser Briwwi. Deel Leit saage aa Abdritt. Duhne
mer des noch yuuse? Gewiss! Guck mol nei. Ya, hoscht
recht; es hot drei.

Vocabulary

    aabaue: to build on (add to)
der Abdritt: privy
die Ax, -e (Ex): axe
    baue: to build
 es Beil, -e: hatchet
 es Biefschteek, -s: beefsteak
der Briehdroog, -dreeg: scalding trough
 es Briwwi, -s: privy
    brode, gebrode: to fry
die Brotwaerscht: sausage
der Brunne: well
die Bump, Bumbe: pump
 es Bumphaus, -heiser: pump house
 es Eise: iron
    eisner: iron (see note)
der Fenseriggel: fence rail

shed is almost empty. We need the wood for the fireplace in the living room and for grandma's wood stove.
The saw, sawbuck, and the chopping block are still in
it.

But we also have a lot of hickory wood for that
smokehouse over there; it is just built right on to the
slaughter house. My granddad always slaughtered himself; <u>his</u> father had built the slaughter house. So
then Dad had two good teachers.

He still uses all the old scrapers, and the old
scalding trough and sausage maker; he and Mom make
liverwurst and (frying) sausage just like years ago.
And he smokes the ham just with hickory wood. On this
(here) farm we have the best bacon and summer sausage
in the county. The beef steaks aren't bad either, I
can tell you.

Now I am hungry. Come on, we'll go back into the
house. Maybe Mom will give us homemade scrapple or
souse for lunch.

That little house over there? Well, you don't know?
That is our privy. Some people say (a synonym). Do
we still use it? Of course! Take a look in (once).
Yes, you're right; it has three.

die Gaardefens, -e: garden fence
    gaerdle: to garden
die Graasmaschien, -e: lawnmower
    graawe, gegraawe: to dig
 es Groossdaadihaus: addition to farmhouse for grand-
      parents
der Grundkeller: ground cellar
der Hackglotz, -gletz: chopping block
    heemgemacht: homemade
 es Heisel: little house
 es Hickerihols: hickory wood
 es Holsheisel: woodshed
der Holskarreb: basket for carrying wood
die Lewwerwa(er)scht or -e: liverwurst
    neibumbe: to pump in(to)
    neischtelle: to place in(to)
der Pannhaas: scrapple

160

der Poschde: post
der Reche: rake
    ruffbumbe: to pump up
die Sache (pl.): things
der Schaawer: scraper
die Schaufel, Schaufle: shovel
die Schipp, Schibbe: spade
    schlachde: to slaughter
 es Schlachthaus, -heiser: slaughter house
    schleife, gschliffe: grind (sharpen)
der Schleifschtee: grindstone
    schmooke: to smoke
 es Schmookhaus, -heiser: smokehouse
der Schupp, Schubbe: shed
der Schpeck: bacon
die Schpring: spring (water)
 es Schpringhaus, -heiser: spring house
der Schubbkarrich, -e: wheelbarrow
 es Schunkefleesch: ham
die Seeg, -e: saw
der Seegbock, -beck: saw horse
    selwer(t): self (oneself)
die Sens, -e: scythe
die Sichel, Sichle: sickle
die Summerwaerscht: summer sausage
der Waerschder (Waschder): sausage maker
    weck (weg): away; weit weck: far away
    wecknemme, weckgenumme: to take away
    weissle: to whitewash
 es Weschheisel: wash house
der Zidderli: souse

Vocabulary Notes

Starting with this chapter, vocabularies will no longer list all plural forms in their entirety. Nouns
whose plurals are formed merely by adding an ending
will be entered thus: die Sens, -e. Such an entry
means that in order to form the plural of the singular Sens, we add the ending -e to get Sense. If
the stem vowel or a final consonant changes to form
the plural, then the entire plural form will be
given: es Haus, Heiser; die Schipp, Schibbe.

<u>schleife, gschliffe</u>: Starting with this lesson, all strong verbs will be followed by their past participles. All weak verbs whose past participles are not formed exactly as described in the grammar text of Chapter Seven will also be followed by their past participles.

<u>eisner</u>: Like <u>schee</u> and <u>glee</u>, <u>eis</u> adds an <u>n</u> before endings.

<u>es Holsheisel</u>: also <u>der Holsschupp</u>, <u>Holsschopp</u>.

<u>die Schpring</u>: also <u>die Gwell</u>.

<u>Grammar</u>

A. Past Participles Used as Adjectives

1. Past participles may be used as adjectives, both predicate adjectives and attributive adjectives. No changes are made and no endings are added to past participles used as predicate adjectives:

Der Schpeck iss   The bacon is smoked.
gschmookt.
Die Oier sin gebrode.   The eggs are fried.

2. However, the <u>weak</u> declension calls for the ending -<u>e</u> in the nominative and accusative cases, singular, where one would normally have no endings:

Der gschmookde Schpeck.. The smoked bacon...
Es gebrodne Oi...   The fried egg...

Note that the <u>e</u> ending of the strong past participle preceding <u>Oi</u> has been dropped, and an <u>n</u> has been added before the adjectival ending <u>e</u>: <u>Es Oi iss gebrode</u>, but <u>es gebrodne Oi</u>. (In a sense, the PG speaker is "putting back" an <u>n</u> that has been dropped from the Standard German past participial ending -<u>en</u>: StG <u>gebraten</u>, PG <u>gebrode</u>. Remember PG <u>glee, gleene</u>, StG <u>klein, kleine</u>; PG <u>schee, scheene</u>, StG <u>schön, schöne</u>.)

3. Past participles use the same <u>mixed</u> and <u>strong</u> declension as do the other adjectives: <u>en geschmookder Schunke, en gebrodnes Oi, ken gebrodne Oier</u>; <u>geschmookder Schunke, gebrodne Oier</u>.

B. Verbs

1. We have seen that the E present perfect is formed by combining the present tense of the auxiliary "have" with the past participle of the main verb (I have bought, he has gone), and that PG follows suit but employs both hawwe and sei as auxiliaries (Ich hab gekaaft, er iss gange).

The E past perfect is formed by combining the past tense of the auxiliary "have" with the past participle of the main verb (I had bought, he had gone). The PG past perfect does not follow suit; it is formed by adding the past participle of the auxiliary to the present perfect tense:

Present Perfect	Ich hab kaaft.
Past Perfect	Ich hab kaaft ghatt (ghadde).
Present Perfect	Er iss gange.
Past Perfect	Er iss gange gwest (gewest).

However, both E and PG past perfect tenses are used for the same purpose, to render action that occurs prior to action expressed in the E past tense or PG present perfect. For example, in the reading selection "Der Hof," the parents placed (past tense) the old wood burning stove into the new summer kitchen. Prior to that, the stove had stood (past perfect) in the kitchen of the main house.

Mei Eldre hen der Holsoffe in die Summerkich geschtellt.	My parents placed the wood stove into the summer kitchen.
Der Offe hot mol in der Kich gschtanne ghatt.	The stove had once stood in the kitchen.

In the second reading selection, the boy wanted to climb (past tense) up into the hayloft, but prior to then, John had never before crawled (past perfect) around a barn.

Ich hab nuffschteigge wolle.	I wanted to climb up.
Der Tschan iss noch nie in re Scheier rumgegraddelt gwest.	John had never before crawled around a barn.

CHAPTER EIGHT

Part Two

Certain buildings are likely to be concentrated near the farmhouse, others are likely to be found near the barn. Again, the following is not necessarily what is, but merely what could be.

Die Scheier waar en Schweizerscheier,
    Ee Dreschdenn un zwee Baare;
En Vorbau un'n Owwerdenn--
Un ich weess aa noch, was mer hen
    Ins alt Dreschdenn neigfaahre;
Flax un Hawwer, weess ich, hen
Mer alsfatt datt uffs Owwerdenn.

Un wann sie waar voll Frucht un Hoi,
    Un aa noch Schteck im Feld,
Dann waare mer die reichschte Leit
Uff daerre aller reichschte Seit,
    Vun daerre ganse Welt;
Der Bauer iss der eensich Mann,
Die Welt allee aadreiwe kann.[1]

[1] H.L. Fischer, from 'S Alt Marik-Haus.

Der Scheierhof

Nooch em Middaagsesse sin mer widder naus. Ich hab
em Tschan unser groossi Scheier weise wolle. Mer sin
am Waageschupp verbeigeloffe. Die verschiddliche Waage
fer die Aerwet uff re Bauerei schtehne datt drin, un aa
en Boggi un en Schpringwaage. Der Paepp hot nadierlich
en Draeckder; er hot die Schaffgeil schunn lang zrick
verkaaft. Alsemol dutt er aa sei Maschien datt nei.

Noh laafe mer die Scheierbrick nuff un schtehne vor
em Scheierdor. Mei Urgroossdaadi hot die aerscht
Scheier do gebaut ghatt; die do iss die zwett. Der
Blitz hot neigeschlaage ghatt. Die Feierleit sin zu
schpeet kumme, un hen gaar nix duh kenne. Die Scheier
iss gans nunnergebrennt. Mei Groossdaadi un sei Paepp
hen graad an sellem Daag es Hoi in die Owwerdenn nuff
gebrocht ghatt. Es Hoi waar nadierlich verlore, awwer
sie hen die Kieh un die Geil nausyaage un nausfiehre
kenne. Paar Woche schpeeder hot meim Groossdaadi sei
Paepp en nei-i Scheier baue losse. Un do schteht sie
noch.

Mer sin in die Dreschdenn neigange. Do hen die Bau-
eremaschiene gschtanne, un hoch owwe, hiwwe un driwwe
uff yeddere Seit, hen mer es Hoi in de Baarre gsehne.
Zwee Leedere hen aa do gschtanne. Ich hab nuffschteige
wolle, awwer der Tschan iss noch nie in re Scheier rumgegraddelt gwest, un er wollt net.

Noh sin mer nunner zum Geilsschtall un Kiehschtall.
Mei Gremmpaepp hot mol viel Kieh ghatt, awwer nau hen
mer yuscht drei. Die gewwe genunk Millich un Raahm fer
uns Kinner. Efders mache mer Kees un drehe (schtoosse)
unser eegeni Budder imme alde Budderfass. Manchmol
bhalde mer die Kelwer un manchmol verkaafe mer sie.

Em Paepp sei Schaffgeil waare arrig scheene Maetschgeil; er hot awwer en Draeckder kaaft un hot die Geil
verkaaft. Awwer mer hen drei gude Reitgeil un en Hutschel fer die Gleene.

Mer gehne darrich der Fuddergang un naus in der
Scheierhof. Noh schtehne mer unner em Vorschuss. Es
hot en paar Daag net gereggert un der Hof iss zimmlich
drucke. Mer kenne niwwer zu der Schteemauer laafe.
Mei Groossdaadi hot die Mauer selwer gebaut mit Schtee

The Farmyard

After lunch we went out again. I wanted to show
John our big barn. We walked past the wagon shed. The
various wagons for the work on a farm stand in there,
and also a buggy and a spring wagon. Father has naturally a tractor; he sold the work horses long ago already. Sometimes he puts his car in there also.

Then we walk up the approach to the threshing floor
and stand in front of the barn door. My great grandfather had built the first barn here; this one is the
second. Lightning had struck. The firemen came too
late, and could do nothing. The barn burned down completely. My grandfather and his father had just
brought on that day the hay into the mow. The hay was
naturally lost, but they could chase out and lead out
the cows and the horses. A few weeks later my grandfather's father had a new barn built. And here it
still stands.

We went into the threshing floor. Here the farm
machinery stood, and high above, on this side and on
that side, we saw the hay in the haylofts. Two ladders
also stood there. I wanted to climb up, but John had
never before crawled around a barn, and he didn't want
to.

Then we went down to the horse stable and cow stable
(stall). My grandfather once had many cows, but now we
have just three. They give enough milk and cream for
us children. Often we make cheese and churn our own
butter in an old churn. Sometimes we keep the calves
and sometimes we sell them.

Dad's workhorses were very nice matched horses; but
he bought a tractor and sold the horses. But we have
three good riding horses and a colt for the little
ones.

We go through the feed corridor and out into the
barnyard. Then we stand under the overhang. It hasn't
rained for a few days and the yard is quite dry. We
can walk over to the stone wall. My grandfather built
the wall himself with stones from the fields. It is

vun de Felder. Sie iss vier bis fimf Fuuss hoch un
geht fascht gans um der Scheierhof rum.

Es watt schpot un der Tschan meecht gaern 's Hinkelhaus sehne. Mer gehne am Seischtall verbei, wu
unser fimf Sei sin. Eeni grickt ball Yunge. Mei
gleene Gschwischder schpiele gaern mit denne Seiche.

Un mer gehne am Hoischtock un aa an der Welschkanngribb verbei, un do simmer. 'S Hinkelhaus iss net
abbaddich grooss. Viel Hinkel sin haus un scharre
eifrich fer Hinkelfudder. Deel Hinkel sin drin un
sitze uff ihre Ruuschde, ihre Hinkelschtange, odder
uff ihre Neschder.

Uff eemol heere mer en Hinkel gaxe. Es schteht vun
seim Nescht uff, un do leit en Oi. Ich heb es uff un
nemm es mit zrick ins Haus. Nix schmeckt so gut wie
frischgelegde Oier un meim Paepp sei gschmokder Schpeck
odder Schunke.

Un marriye frieh, wann der Haahne greeht, schteh
ich uff un mach mer gebrodne Oier un Schunkefleesch.

Vocabulary

die Aerwet, Aerwedde: work
der Baarre: loft of barn
die Baueremaschien, -e: farm machine
 es (die) Boggi, -s: buggy
    bringe, gebrocht: to bring
der Budder: butter
 es Budderfass, -fesser: butter churn
der Draeckder, -s: tractor
    drehe: to churn butter, to turn
die Dreschdenn: threshing floor
    eegen: own
    efders: often
    eifrich: diligently, busily
der Feiermann, -leit: fireman
    frischgelegt: freshly laid
der Fuddergang, -geng: passageway along stalls to
        facilitate feeding
der Fuuss: foot (unit of measure)
der Gaul, Geil: horse
    gaxe: to cackle

four to five feet high and goes almost entirely around
the barnyard.

It is getting late and John would like to see the
chicken house. We go past the pigsty where our five
pigs are. One will soon have young ones. My little
brothers and sisters like to play with those piglets.

And we go past the haystack and also the corn crib,
and here we are. The chicken house is not particularly large. Some chickens are outside and are scratch-
ing busily for chicken feed. Some chickens are inside
and are sitting on their roosts, (synonym), or on
their nests.

All of a sudden we hear a chicken cackling. It gets
up from its nest, and there lies an egg. I pick it up
and take it along into the house. Nothing tastes so
good as freshly laid eggs and my father's smoked bacon
or ham.

And tomorrow morning, when the cock crows, I'll get
up and make myself fried eggs and ham.

gebrode: fried
der Geilsschtall, -schtell: horse stall, stable
die Gleene (pl.): the little ones
gschmookder: smoked
es Hinkel: chicken
es Hinkelfudder: chicken feed
es Hinkelhaus: chicken house
die Hinkelschtang, -e: chicken roost
hiwwe: on this side
es Hoi: hay
der Hoischtock, -schteck: hay stack
es Hutschel, -cher: colt
der Kees: cheese
der Kiehschtall, -schtell: cow stall, stable
die Kuh, Kieh: cow
die Leeder, -e: ladder
die Maetschgeil (pl.): matched horses
die Maschien, -s: car
die Mauer, -e: wall
mitnemme, mitgenumme: to take along
nausyaage: to chase out

neischlaage, neigschlaage: to strike (lightning)
es Nescht, Neschder: nest
nuffschteige, nuffgschtigge: to climb up
nunnerbrenne, nunnergebrennt: to burn down
es Oi, -er: egg
die (es) Owwerdenn: hayloft
der Raahm: cream
der Reitgaul, -geil: riding horse
schaare (schaerre): to scratch
die Scheierbrick: approach to threshing floor of a
    bank barn
es Scheierdor: barn door
der Scheierhof, -heef: barnyard
schmecke: to taste
der Schpringwaage, -wegge: spring wagon
die Schteemauer, -e: stone wall
schtoosse, gschtoosse: to churn butter; to bump,
    push
es Seiche, -r: piglet
der Seischtall, -schtell: pigpen
uffhewe, uffghowe: to pick up; to lift up
verkaafe: to sell
verliere, verlore: to lose
verschiddlich (verschiedlich): various
der Vorschuss: overhang of a Swiss barn on barnyard
    side
der Waage, Weege (Wegge): wagon
die Welschkanngribb: corn crib
die Yunge (pl.): young (of animals)

Vocabulary Notes

der Baarre: Other terms that can be used for the loft
  of a barn are der Hoibaarre, der Fruchtbaarre. Die
  Owwerdenn is the loft area immediately above the
  threshing floor, die Dreschdenn.

die Dreschdenn: Other terms are die Dreschdennd, der
  Dreschfloohr, Dreschflaar, die Scheierdenn.

es Hinkelhaus: Heisel can be used just about every
  time Haus refers to a relatively small building:
  Hinkelhaus, Hinkelheisel; Holshaus, Holsheisel; as
  well as Weschheisel, Schmookheisel, Schlachtheisel,

and the like. Chickens are also kept in der Hinkel-
kewwich or Hinkelbenn.

es Hutschel: Also es Hutsch, Hutschelche, Hutschli.

die Welschkanngribb: Some other terms are die Welsch-
kannkribb, es Welschkannheisel, es Welschkannhaus,
es Welschkannheisli.

es Seiche: Also Seili, Wutzel, Wutzli, Witzel, Wit-
zelche.

CHAPTER NINE

Part One

In this lesson our young guide will take us through
the garden and fields of his family's farm. Watch for
infinitives and infinitive phrases, some of which are
accompanied by zu, some of which are not. The vocabulary is once again rather extensive, but the grammar is
relatively short.

Sell waar mei Mudder ihr Gaarde,
Vun Ungraut immer frei,
Un all die Lenner waarde
Uff ihre Blanserei.
Es sin etliche gleene Greewe
Mit Zwiwwlesuume gseet,
Des Friehselaatschtick newe
Wu owwe der Buchsschtock schteht.

En Aerbiereschtick, un Gwendel
Blobarriger- un Salwei Tee,
Un Saffrich mit geelrode Bendel
Duhne aa im Gaarde schteh.
'S gebt aa Andiffde un Riewe,
Un Yuddekaersche fer Pei,
Un hie un dadde schiewe
Paar fremme Blanse sich nei.[1]

[1]*Michael A. Gruber, from "Mei Mudder Ihre Gaarde,"*
PG Verse.

171

Die Felder; Es Aerscht Deel

Ich weess, du witt nunner zum Schwamm un naus in die
Felder, awwer mer gehne mol in unser Gaarde nei. Mer
gleiche all zu gaerdle, abbaddich mei Memm un mei
Groosseldre. Der Gremmpaepp kann net waarde, im Friehyaahr im Gaarde zu graawe. Dann hot er der gans Summer
viel Aerwet zu duh.

Mer kenne do darrich des glee Dierli nei. Ya, mer
muss do uff em Land en Gaarde eifense. Weescht, weeich
de Haase un Hinkel. Die fresse immer die yunge Blanse.
Sie gleiche Gaardesach, abbaddich Sellaat, Sellerich,
un Schpinaat. Un die Fens iss aa, fer die Kinner draus-
zuhalde. Sie gleiche do in der Gaarde zu kumme, fer
die Reddiche un Geelriewe aus em Grund zu robbe un zu
esse. Un die Tomaets gleiche sie aa abzurobbe.

Un do sin gleene Grautkepp. Die misse aerscht groos
warre, dann macht die Memm Sauergraut. Die Gummere do
mache aa guder Gummeresellaat. Des sin Zwiwwle, die sin
Aerbse, un selli sin Rotriewe. Ya, heemgereest Gemies
iss awwer ebbes Gudes.

Selli? Selli sin doch Grummbeereschtengel. Was
meenscht, wu sin die Grummbeere? Die sin doch unne im
Grund, wu die Watzle waxe! Mer muss en Schpaat hawwe
fer sie auszuschteche odder auszugraawe. Draus im Feld
yuuse mer en Grummbeereausmacher, awwer net do im
Gaarde.

Ya, ich weess, des iss Umgraut. Marriye muss ich
die Gaardehack nemme, un draageh. Ich duh awwer net
gaern im Gaarde hacke. Sell iss zu hatt zu duh.

Gleicht unser Butzemann? Mer hen en datthie-
gschtellt, fer die Grabbe abzuschrecke. Awwer der alt
Butzemann muss denke, es iss gaar net schee, die Veggel
zu schrecke. Ich sehn en nie so ebbes duh.

Uff der anner Seit der Scheier iss der Schwamm.
Mannichmol saagt mer aa Baschdert dezu. Datt drunne
iss es fascht immer nass, abbaddich im Friehyaahr. En
Bechli fliesst do darrich un datt in die Grick nei.
Datt unnich denne Weide(beem) iss es immer schee kiehl,
awwer mer hen ken Schtiwwel aa; mer gehne liewer nuff
in die Wiss un noch de Felder. Es iss viel druckner
datt drowwe. Guck mol die Fleddermeis!

The Fields; Part One
_____

I know you want to go down to the meadow and out into the fields, but we will first go into our garden.
We all like to garden, especially my mom and my grandparents. Granddad can't wait to dig in the garden in
the spring. Then he has a lot of work to do all summer.

We can go through this little gate. Yes, one must
fence in a garden here in the country. You know, because of the rabbits and chickens. They always eat the
young plants. They like garden truck, especially the
lettuce, celery, and spinach. And the fence is also to
keep out the children. They like to come into the garden and pull the radishes and carrots out of the ground
and eat them. And they also like to pick the tomatoes.

And here are little cabbage heads. They must first
get big, then Ma makes sauerkraut. These cucumbers
here also make good cucumber salad. These are onions,
these are peas, and those are red beets. Yes, home
raised vegetables are something good.

Those? Those are potato stalks. What do you mean,
where are the potatoes? They are down in the ground,
where the roots grow! One must have a spade to dig
them out or (synonym). Out in the field we use a potato
digger, but not here in the garden.

Yes, I know, these are weeds. Tomorrow I must take
a garden hoe and get to it. But I don't like to hoe in
the garden. That is too hard to do.

Do you like our scarecrow? We put him there to
scare off the crows. But the old scarecrow must think
it isn't very nice at all to scare the birds. I never
see him doing anything like that.

On the other side of the barn is the wet meadow.
Sometimes one also says Baschdert for it. Down there
it is almost always wet, especially in the spring. A
brooklet runs through it and there into the creek.
There under those willow (trees) it is always nice and
cool, but we don't have any boots on; it's better that
we go up into the meadow and to the fields. It is much
drier up there. Just look at those butterflies!

Do howwe waxt aa gut Hoi. Ver alders hen die Bauere
es Hoi mit ihre deitsche Sense abgemeeht. Mei Gremmpaepp hot noch sei deitschi Sens un Dengelschtock. Er
lannt mich mol dengle, awwer ich muss aerscht elder
warre. Heit hen mer en Meehmaschien, fer es Hoi zu
meehe. Un ver alders hen sie es Hoi mit re Hoigawwel
uff der Hoiwaage, der Leederwaage, gegawwelt, hen es in
die Scheier gebrocht, un hen es in der Baarre nuffge-
gawwelt. Heidesdaags kummt es Hoi aus re Maschien gans
schee uffgebunne mit Droht odder Schnur. Sehnscht die
viel Hoischrecke do rumhupse?

Un datt uff der annere Seit der Heckefens iss unser
Baamgaarde, odder Bungert. Do howwe hen mer verschiddliche Sadde Ebbel fer Schnitz, Ebbelboi, Lattwarrick,
un Seider zu mache. Mer hen aa en paar Beerebeem. Die
Kaschebeem un Paschingbeem sin unne beim Gaarde. Die
Ebbel sin nadierlich net reif bis im Schpotyaahr. Dann
kumme mir Kinner do haer un robbe en Abbel ab un beisse
nei. Die Brieh schpritzt graad so raus. En frischer
Abbel schmeckt awwer gut.

Vocabulary

aahawwe, aaghatt: to have on, wear (clothes)
der Abbel, Ebbel: apple
abmeehe: to mow down
abrobbe: to pick off, to pluck
abschrecke, abgschrocke: to scare off
die Aerbs, -e: pea
ausgraawe, ausgegraawe: to dig out
ausschteche, ausgschtoche: to dig out (as potatoes)
der Baamgaarde: orchard
der Baschdert: wet field used for pasture
 es Bechli, -n: brook
der Beerebaam (Bierebaam), -beem: pear tree
beisse, gebisse: to bite
die Blans, -e: plant
die Brieh: juice
der Bungert: orchard
der Butzemann, -menner: scarecrow
datthieschtelle: to place (stand) over there
der Dengelschtock, -schteck: anvil upon which scythe
     is sharpened by hammering
dengle, gedengelt: to sharpen by hammering

Good hay also grows up here. Long ago the farmers mowed the hay with their German scythes. My granddad still has his German scythe and hammering anvil. He's going to teach me to "hammer" some time, but I must first get older. Today we have a mowing machine to mow the hay. And long ago they forked the hay onto the hay wagon, the ladder wagon, with a hay fork, brought it into the barn and forked it up into the mow. These days the hay comes out of a machine very nicely tied up with wire or string. Do you see the many grasshoppers jumping around here?

And there on the other side of the brush fence is our orchard, or (synonym). Up here we have various kinds of apples to make snitz, apple pie, apple butter, and cider. We also have a few pear trees. The cherry trees and the peach trees are below at the garden. The apples are naturally not ripe till fall. Then we children come here and pluck an apple off and bite into it. The juice just squirts out. My, but a fresh apple tastes good.

es Dierli, -n: gate
    draageh, draagange: to begin doing something, to
        start in
    draushalde, drausghalde: to keep someone or some-
        thing out
der Droht, Dreht: wire
der Ebbelboi, -s: apple pie
    eifense: to fence in
die Fleddermaus, -meis: butterfly
    fliesse, gflosse: to flow
    fresse, gfresse: to eat (said of animals)
    gaardegraawe: to dig in the garden
die Gaardehack, -e: garden hoe
 es Gaardesach, -e: garden produce
    gaww(e)le: to fork, to pitch (as with a hay fork)
die Geelrieb, -riewe: carrot
 es Gemies: vegetables
die Grabb, -e: crow
der Grautkopp, -kepp: cabbage head
die Grick, -e: creek
die Grummbeer (-bier), -e: potato
der Grummbeereausmacher: potato digger

der Grummbeereschtengel, -schtengle: potato stalk
die Gummer, Gumm(e)re: cucumber
der Gummeresellaat, -e: cucumber salad
die Heckefens, -e: brush fence
    heemgereest: home raised
    heidesdaags: these days
die Hoigawwel, -gawwle: hay fork
der Hoischreck, -e: grasshopper
der Hoiwaage, -wegge: hay wagon
der Kaschebaam (Kaerschebaam), -beem: cherry tree
der Lattwarrick (Lattwaerrick): apple butter
der Leederwaage, -wegge: (hay) wagon with rack
die Maschien, -e: machine
    meehe: to mow
die Meehmaschien, -e: mower
    neibeisse, neigebisse: to bite into (something)
    nuffgawwle, nuffgegawwelt: to fork up, pitch up
der Paschingbaam (Paerschingbaam), -beem: peach tree
    rausschpritze: to squirt out
    reif: ripe, mature
    robbe: to pick, pluck
die Rotrieb, -riewe: red beet
der Sauergraut: sauerkraut
die Schnitz (pl.): dried sections of apple, dried fruit
der Schpaat, Schpaade: spade
der Schpinaat: spinach
    schpritze: to squirt
    schrecke, gschrocke: to scare
der Schtiwwel: boot
der Schwamm, Schwemm: meadow
der Seider: cider
der Sellaat: lettuce, salad
der Sellerich: celery
die Tomaet (Tomat), -s: tomato
    uffbinne, uffgebunne: to tie up
 es Umgraut (Ungraut): weeds
    ver alders: long ago, in former times
die Wa(r)tzel, Wa(r)tzle: root
der Weidebaam, -beem: willow tree
die Wiss, -e: meadow
die Zwiwwel, Zwiwwle: onion

Vocabulary Notes

Aerbs: also Arrebs, -e.

der Baamgaarde: also der Boomgaarde, Bammgaarde, Bummert, Bungert.

dengle: A farmer could take with him a Dengelschtock, a small anvil, and a Dengelhammer, a small hammer, so that if his German scythe became dull while he was working in the field, he could pause and "dengel" the scythe sharp again by placing it on the anvil and tapping it with the hammer to draw out the metal edge to a fine sharpness that rivalled whetting or grinding.

heidesdaags: also heidzudaags, heidzedaags, heidichdaags.

der Hoischreck: also der Hoischrecker.

die Meehmaschien: also die Graasmaschien, der Rieber, der Graasrieber.

reif: also zeidich.

schmecke: also schmacke, but for some PG speakers schmacke means "to smell."

Grammar

A. Infinitives

1. In E, we many times have to use to with an infinitive:

He has to go into town.
He is allowed to buy some candy.

The verbs "go" and "buy" must be preceded by to.

Other times we must not use to:

He must go into town.
His parents let him buy candy.

After modals and after verbs like "let," we must never use to.

2. Much the same situation exists in PG. As we have

noted since Chapter One, <u>zu</u> should never be used with infinitives after modals and <u>brauche</u>:

Der Gremmpaepp kann net waarde.	Granddad can't wait.
Mer muss en Schpaat hawwe.	One must have a spade.
Er brauch des net duh.	He doesn't have to do that.

We have also noted that <u>zu</u> is not used with infinitives dependent on certain verbs: <u>duh</u> "to do," <u>heere</u> "to hear," <u>sehne</u> "to see," <u>helfe</u> "to help," <u>losse</u> "to let, allow," and <u>lanne</u> "to teach, learn." Two other such verbs have not yet been used in our reading selections: <u>fiehle</u> "to feel," and <u>heesse</u> "to request, bid."

The following are some examples taken from this lesson:

Ich duh net gaern hacke.	I don't like to hoe.
Ich sehn en nie so ebbes duh.	I never see him doing something like that.
Mei Brieder helfe em Paepp dresche.	My brothers help my father thresh.
Sehnscht die Hoi-schrecke rumhupse?	Do you see the grasshoppers jumping around?
Er lannt mich mol dengle.	Sometime he's going to teach me to "sharpen" the scythe.

3. However, the infinitive is used with <u>zu</u> when it modifies

(a) A noun

Gremmpaepp hot viel Aerwet zu duh.	Granddad has a lot of work to do.

(b) An adjective

Sell iss hatt zu duh.	That is hard to do.
Es iss net schee, die Veggel zu schrecke.	It isn't nice to scare the birds.

4. Purpose is expressed in PG with an infinitive phrase beginning with <u>fer</u> and ending with the infinitive. Actually, the <u>zu</u> may be omitted before the infinitive, but it is used consistently in this lesson

to make identification of these infinitive phrases easier.

..., fer Reddiche zu esse	..., (in order) to eat radishes
..., fer es Hoi zu meehe	..., to mow the hay

Note that the <u>zu</u> is "sandwiched" between the separable prefix and verb:

..., fer die Kinner drausszuhalde	..., to keep the children out
..., fer sie auszugraawe	..., to dig them out

5. Note also that the normal E translation of such a PG infinitive phrase <u>begins</u> with the infinitive, then goes back to the beginning of the PG phrase and follows through to the end.

6. One often finds <u>ze</u> instead of <u>zu</u> in PG literature.

Part Two

If you studied the infinitive structures presented
in the first part of this lesson, you are ready to proceed with no trouble. In this section you will en-
counter nothing new in grammar, but you may want to add
some of the new words to your "active" vocabulary.

*Wammer alleweil in unsrer Gegend rumkummt
un sicht die heerliche Felder un Wisse, die
scheene Wuhnheiser un brechdiche groosse
Scheire, mit allerhand hendiche Newgebei un
vielerlee Maschiene, kann mer schier net
glaawe, dass es net so gwest waar fer Alders.
Schtatt unsre gude Blick, Eege, un Wallse,
hen sie so Blick mit hilsne Wennbredder
ghatt, wu sie's Land mit rumgschunne hen,
Eege mit hilsne Zappe, odder hen yuscht en
Ascht vumme Baam iwwers Feld gschleeft, wann
sie gseet hen ghatt; sunscht iss als mit der
Hack un vun Hand gschafft warre. Mer brauch
yuscht an die alt Sichel un die Sens zu
denke, un die Maschiene mit vergleiche, um
sich zu verwunnere.*[1]

[1]*A.R. Horne, from "Vor Alders,"* Pennsylvania German
Manual.

181

Die Felder; Es Zwett Deel

Un uff der annre Seit sellre Schteefens iss unser
Holsland. Mer graddle mol do driwwer. Ei! Hoscht des
gsehne? En Schlang lenger ass mei Aar(e)m. Nee, 's
waar ken Rasselschlang. Ich hab ken Rassle gsehne. En
schwazi Schlang verleicht. Die Schlange wuhne do
zwische de Schtee. Ich denk, sie esse die Wissemeis un
verleicht aa die Fensemeis.

Mer hen schunn mannicher Glofder Hols do rausgenumme.
Eechehols un Maebelhols brenne es bescht im Feierheerd;
sie sin alle beed arrig hatt un brenne schee langsam.
Awwer wie gsaa(g)t, mer yuuse yuscht Hickerihols fer es
Schunkefleesch zu schmooke. Baerge (Baergge) finnscht
aa iwwerall im Busch, awwer Baergehols iss zu weech, un
es brennt zu schnell. Un die Bein(d)beem un Schpruuss
sin iwwerhaabt zu weech, fer Brennhols gut zu sei.
Halt mol! Datt sin zwee Gschwalls am Schpiele, odder
verleicht will der eent em annere sei Nuss odder Eechel
wecknemme.

Nau gehne mer darrich der Busch. Uff der annere
Seit em Busch sin meh Felder. Ya, hoscht recht, des
iss en Schpring(s)eeg. Aerscht bluugt mer, un dann
brauch mer en Eeg, fer die groosse Scholle uffzubreche.
Do simmer. Mer reese Welschkann, Weeze, Rogge, un Hawwer. Nee, es Welschkann hot noch ken Kolwe. Die
Schtengel misse noch meh waxe, un heecher warre. Guck
mol datt drowwe--en Hinkelwoi.

Awwer der Weeze sehnt schunn gut aus. Mer hen es
Weezefeld schunn letscht Schpotyaahr gseet. Er watt
ball zeidich, un dann bringe mer die Dreschmaschien
ruff un dresche es gans Feld. Mei eldere Brieder helfe
em Paepp dresche, un des geht gans schnell. Die Aern
geht schneller mit re Maschien.

Ich glaab, es watt wennich schpot. Mer gehne nunner
zum Deich un dann zrick zum Haus. Unser Deich iss net
groooss un net arrig dief. Awwer mer kumme immer im
Summer do raus un gehne schwimme. Un im Winder gehne
mer schkeede. Gewiss, es Eis iss immer dick genunk,
fer schkeede geh. Do sin aa gleene Fisch do drin im
Wasser un viel Fresch. Un mer sehnt immer Gens un Ende
rumschwimme. Alsemol sehnt mer aa Schillgrodde un
Muschkradde.

The Fields; Part Two

And on the other side of that stone fence is our
woodland. Let's crawl over it here. Hey! Did you see
that? A snake longer than my arm. No, it wasn't a
rattlesnake. I didn't see any rattles. A black snake,
perhaps. The snakes live here between the stones. I
think they eat the meadow mice and perhaps also the
chipmunks.

We've taken many a cord of wood out of there. Oak
wood and maple wood burn the best in the fireplace;
they are both very hard and burn nice and slow. But as
I've said, we use just hickory wood to smoke the ham.
Birch you also find everywhere in the woods, but birch
wood is too soft, and it burns too quickly. And the
pine trees and spruce are especially too soft to be
good for firewood. Hold it! There are two squirrels
playing, or maybe the one wants to take away from the
other his nut or acorn.

Now we'll go through the woods. On the other side
of the woods are more fields. Yes, you are right, that
is a spring harrow. First one plows, and then one
needs a harrow to break up the big clumps of earth.
Here we are. We raise corn, wheat, rye, and oats. No,
the corn doesn't have any ears yet. The stalks have to
grow more, and get higher. Take a look up there--a
chicken hawk.

But the wheat looks good already. We sowed the
wheatfield last fall already. It will get ripe (mature)
soon, and then we'll bring the threshing machine up and
thresh the whole field. My older brothers help my
father thresh, and that goes very quickly. The harvest
goes more quickly with a machine.

I think it's getting a little late. We'll go down
to the pond and then back to the house. Our pond is
not big and not very deep. But we come out here in the
summer and go swimming. And in winter we go skating.
Certainly, the ice is always thick enough to go skating.
There are also little fish in there in the water and
many frogs. And one always sees geese and ducks swimming around. Sometimes one also sees turtles and musk-
rats.

Un die Schtadtleit denke, sie hen es schee.

Kumm, mer gehne zrick. Ich wa mied, un du verleicht
aa. Datt iss der Gremmpaepp widder im Gaarde; er iss
am Wessre (Wessere). Er guckt immer niwwer zu der
Drauwerank. Er kann gaar net waarde, bis die Drauwe
zeidiche. Sei Weifass iss beinaah leer. Ich ess sie
awwer liewer. So en groossi, blo-i Draub iss ebbes
Gudes.

Un die Hembeere sin aa gut--die rode un die schwaze.
Dann wann sie un es anner Obscht zeidiche, do sin sie--
die Memm, Gremmemm, un Schweschdere--daagelang am Eimache. Dann hen mer awwer der gans Winder Obscht uff
em Disch. Un nau wessert mir es Maul. Ich geh mol nei
un froog die Memm, wann mer esse.

Vocabulary

der Aar(e)m, -e: arm
die Aern: harvest
die Baerge (Baergge): birch
 es Baergehols: birch wood
    bluuge: to plow
    daagelang: for days (at a time)
    dief: deep
die Draub, Drauwe: grape
die Drauwerank, -e: grape vine
    dresche: to thresh
die Dreschmaschien, -e: threshing machine
der Eeche: oak
 es Eechehols: oak wood
 es Eechel: acorn
die Eeg, Eege: harrow
    eimache: to preserve (e.g., fruit)
die Ent, Ende: duck
die Fensemaus, -meis: chipmunk
    finne, gfunne: to find
der Fisch: fish
der Frosch, Fresch: frog
 es Glofder: cord of wood
    graddle, gegraddelt: crawl
der Gschwall, -s: squirrel
    halde: hold; stop
der Hawwer: oats

And the city people think they have it nice.

Come, we're going back. I'm getting tired and maybe you too. There is Granddad in the garden again;
he's watering. He always looks over at the grape vine.
He just can't wait till the grapes get ripe. His wine
barrel is almost empty. But I rather eat them. Such
a big, blue grape is something good.

And the raspberries are also good--the red ones and
black. Then when they and the other fruit ripen, they
--Mom, Grandma, and sisters--are preserving fruit for
days. But then we have fruit on the table the whole
winter. Now my mouth is watering. I'm going in and
ask Mom when we eat.

die Hembeer, -e: raspberry
der Hinkelwoi: chicken hawk
 es Holsland: woodland
    iwwerhaabt: especially
der Kolwe: ear of corn, cob
    leer: empty
 es Maebelhols: maple wood
 es Maul: mouth
    mied: tired
die Muschkratt, -radde: muskrat
die Nuss, Niss: nut
die Rassel, Rassle: rattle
die Rasselschlang, -e: rattlesnake
    reese: to raise
der Rogge: rye
die Schill(t)grodd, -e: turtle, tortoise
die Schlang, -e: snake
der Scholle: clod
die Schpring(s)eeg: spring harrow
der Schpruuss: spruce
der Schtee: stone
die Schteefens, -e: stone fence
der Schtengel: stalk
    see-e, gseet: to sow
    uffbreche, uffgebroche: to break up
    weech: soft
der Weeze: wheat

es Weezefeld, -er: wheat field
es Weifass: wine barrel, cask
es Welschkann: corn
   wess(e)re: to water
   wie gsaagt: like I said; as has been stated
die Wissemaus, -meis: meadow mouse
   zeidich: ripe
   zeidiche: to ripen, mature

Vocabulary Notes

der Gschwall: Many variants exist for this particular
   animal: der Schgwall, der Eecher, der Eechhaas, es
   Eechhannche, es Eecherli are just some of them.

der Woi: In some areas this same bird would be called
   der Habich.

CHAPTER TEN

Part One

As our young friend takes us through a typical day, watch for reflexive verbs and pronouns (e.g., I wash myself), and commands (e.g., Sit down!). They are both very important concepts in the Pennsylvania German dialect.

Vor viel Yaahre zrick hen die Bauere viel hedder schaffe misse wie alleweil, un fer sell hen sie aa efders esse misse. Sellemols hot mer nix gewisst vun Mehmaschiene. Es Graas un die Frucht iss all mit der Sens un dem Reff gemeeht warre. Es iss aa vun frieh marriyeds bis schpot oweds gschafft warre. Um fimf iss mer uffgschtanne, hot zu Marriye gesse, un eb die Sunn recht uffwaar, waar mer uffem Feld. Um ebaut nein Uhr iss en Schtick Esse uffs Feld gebrocht warre-- Brot, Kuche, Pei, un so weider, un alsemol aa en bissel Wisski. Sell hot mer es Neinuhrschtick gheese. Dennoh hot mer gschafft bis Middaag. Noochmiddaags hot mer des Vieruhrschtick grickt. Dann hot mer gschafft bis oweds. Mannichmol iss mer nooch dem Sobber widder nausgange un hot gschafft bis dunkel Nacht.[1]

---

[1]*Daniel Miller, from "Lunsch Uffem Feld Un In Der Kaerrich,"* Pennsylvania German *2nd ed.*

En Gewehnlicher Daag; Es Aerscht Deel

Im Summer geht die Sunn zimmlich frieh uff, awwer die aerschde Sunneschtraahle finne mich schunn wacker. Es iss Zeit fer uffzuschteh. Ich bin net eener vun denne Schloffkepp; ich duh mich nie net verschloofe, un brauch aa ken Weckuhr fer mich uffzuwecke.

Mei Eldre, mei eldere Brieder un mei elderi Schweschder sin schunn lang uff. Der Paepp un mei Brieder gehne gewehnlich nunner in die Kich un Summer-kich un wesche sich un balwiere sich datt drunne. Dann kenne mei Memm un mei Schweschder die Baadeschtubb yuuse.

Die sin awwer ball faddich (faerdich), un dann kann ich neigeh. Des nemmt mich aa net lang. Ich kann mich wesche, abdrick(e)le un schtreehle schneller ass eenichebber. Un ich kann mich aa schnell aaduh, abbad-dich im Summer; ich hab yo ken Schul.

Ich kumm graad aus der Baadeschtubb raus, do heer ich schunn die Memm rufe, "Kinner, schteht uff!" Awwer ich muss mei gleener Bruder un gleeni Schweschder wid-der rufe, "Billi, schteh uff! 'S Zeit, Betz; kumm raus un wesch dich!" Dann heer ich die Memm ruffrufe, "Sei schtill datt owwe! Es Bobbel schlooft noch." Alle Marriye.

Drunne hen sie sich schunn an die Aerwet gemacht. Die Mannsleit sin draus in der Scheier un im Hinkel-haus, un die Weibsleit sin in der Kich un richde es Marriye-esse. Un ich muss naus an die Scheier, fer die Kieh melke helfe. Ich nemm en Melkeemer un hock mich uff en Melkschtuhl newe der Kuh. Noh nemm ich der Kiwwel zwische die Bee un fang aa zu melke.

Un do kummt sie schunn. Alle Marriye kummt sie, hockt sich hie, un guckt mich aa. Dann muss ich ihr enwennich frischi Millich in ihre Schissel schidde. Noh drinkt sie die Schissel leer, hockt do en paar Mi-nudde un wescht sich, iss dann zufridde, un geht zrick zu ihre Yunge. Sie iss en gudi Mudder zu ihre Ketzlin.

Ball bin ich faddich, un kann ins Haus, fer mei Breckfescht esse. Die gans Familye sitzt schunn am Disch. Die Groosse esse gebrodne Schunkefleesch un Oier, die Gleene esse ihre Brei odder Oier (die Gleene saage Gacki).

A Typical Day; Part One

In summer the sun rises quite early, but the first
sun rays find me awake already. It is time to get up.
I am not one of those sleepy heads; I never oversleep,
and I also don't need an alarm clock to wake me up.

My parents, my older brothers and my older sister
have already been up for a long time. Dad and my
brothers usually go down into the kitchen and summer
kitchen and wash themselves and shave themselves down
there. Then my mom and my sister can use the bathroom.

But they are soon finished, and then I can go in.
That also doesn't take long. I can wash myself, dry
myself off, and comb myself faster than anyone. And I
can get dressed quickly too, especially in summer; I
don't have any school, you see.

I am just coming out of the bathroom, when I hear
my mother calling already, "Children, get up!" But I
must call my little brother and little sister again,
"Billy, get up! It's time, Betts; come out and wash
yourself." Then I hear Mom calling up, "Be quiet up
there! The baby is still sleeping." Every morning.

Downstairs they have already started in to work.
The men are out in the barn and chicken house, and the
women are in the kitchen and preparing breakfast. And
I must go out to the barn to help milk the cows. I
take a milk pail and sit down on a milking stool beside
the cow. Then I put the pail between my legs and begin
to milk.

And here she comes already. She comes every morning, sits down, and looks at me. Then I must pour a
little fresh milk into her bowl. Then she drinks the
bowl empty, sits there a few minutes and washes herself, is then satisfied, and goes back to her young
ones. She is a good mother to her kittens.

Soon I am finished and can go into the house to eat
my breakfast. The whole family is already sitting at
the table. The adults are eating fried ham and eggs,
the children are eating their porridge or eggs. (The
little ones say ---.)

Demarriye misse mir Kinner net widder nuff, fer uns
zu wesche un unser gude Schulgleeder aaduh, fer uns
uff der Weg noch der Schul mache. Es iss yo Summer,
Gott sei dank. Marriyeds vor der Schul bedrachde sich
un schtreehle sich die Meed schtunnelang im Schpiggel.
Dann aergere mir Buwe uns, verloss dich druff. Ball
greische mer uns nanner aa, un noh ruft die Memm widder zum Schpeicher ruff, "Seid schtill datt owwe!"
Alle Marriye.

Awwer wie gsaagt, es iss Summer, un mer mache uns
widder an die Aerwet. Die Meed misse meinre Mudder im
Haus helfe, un die Buwe misse im Hof un draus in de
Felder mithelfe. Mei gleener Bruder muss die Bortsch
abkehre (abfeege), awwer der Bese(m)schtiehl iss wennich zu lang fer ihn, un er schlaagt immer wedder 's
Haus odder en Fenschder. Noh kann ich die Memm rufe
heere, "Bass uff! Schlaag's Fenschder net nei!" Alle
Marriye.

Un ich muss noch der Scheier, un die Schtell ausmischde. Ich geh awwer aerscht zum Tschimm, unser
Kettehund. Er iss en groosser Hund, viel greesser ass
unser Wuffi, awwer er waar immer wennich blaffich un
nau iss er unne beim Scheierhof mit re lange Kett aagebunne. Do kumme nie net kee Rumleefer (Draemps) rum,
fer sich nachts im Scheier zu verschteck(e)le.

Ich nemm die Mischtgawwel un fang aa, auszumischde.
Do muss der Mischthaahne awwer dabber aus em Weg; der
Mischt fliegt graad so naus uff der Mischthaufe. Ich
kann net saage, ass ich gleich, die Schtell auszumischde. Ich dummel mich.

Dann muss ich aa der Schlabbkiwwel nemme un die Sei
schlabbe. Der Droog soll schee voll sei; dann warre
die Sei schee fett. Noh geh ich noch eemol zum Tschimm,
schpiel wennich met em un geh dann nuff in der Gaarde.

Im Gaarde nemm ich die Giesskann un fang aa, die
Blanse Wasser gewwe. Uff eemol heer ich, "Schitt net
so viel Wasser uff die Blanse, du Dummkopp. Du wescht
sie yo weck!" Ich dreh mich schnell rum, un datt
schteht der Gremmpaepp. "Du brauchscht dich net zu
dummle. Hoscht yo der gans Daag."

Un ich denk, "Schwetz net so dumm," awwer ich saag
es net. Ennihau, ich muss nau im Gaarde uffheere. Es

This morning we children don't have to go up (stairs) again to wash ourselves and put on our good school clothes, to get on our way to school. It is summer, you see, thank God. In the morning before school the girls look at themselves and comb themselves for hours in front of the mirror. Then we boys get angry, you can depend on that. Soon we are shouting at each other and then Mom calls up to the second floor, "Be still, up there!" Every morning.

But as (I've) said, it is summer, and we start in to work again. The girls have to help my mother in the house, and the boys have to help along in the (farm) yard and out in the fields. My little brother must sweep off the porch, but the broom handle is a little too long for him, and he keeps hitting against the house or a window. Then I can hear Mom call, "Watch out! Don't knock in the window!" Every morning.

And I have to go to the barn and clean out the stalls. But first I go to Jim, the dog we keep tied up. He is a big dog, much larger than our Wuffi, but he was always a little "barky," and now he is tied up with a long chain down at the barn. Tramps never come around here to hide in the barn at night.

I take the manure fork and begin to clean out the manure. The (manure) rooster has to get out of the way quickly; the manure just flies right out onto the manure pile. I can't say that I like to clean out the stalls. I hurry (up).

Then I also have to take the slop pail and slop the pigs. The trough should be nice and full; then the pigs get nice and fat. Then I go once more to Jim, play a little with him, and then go up into the garden.

In the garden I take the watering can and begin to give the plants water. Suddenly I hear, "Don't pour so much water on the plants, you dummy. You'll wash them away!" I quickly turn around, and there stands my granddad. "You don't have to hurry. (You) have the whole day."

And I think, "Don't talk so dumb," but I don't say it. Anyhow, I have to stop in the garden now. It's

iss Zeit, fer es Geilgscharr (-gschaerr) hole un en
Gaul zum Weggeli eischpanne. Ich muss es Zehuhrschtick naus zum Paepp un meine Brieder bringe. Die
misse aa ebbes zu drinke un esse hawwe. Dann muss ich
widder der Gaul ausschpanne un es Weggeli widder in
der Waggeschopp weckschtelle.

Noh muss ich noch's Graas im Hof schneide (meehe),
un der Marriye iss verbei. Ich geh ins Haus un finn
es Middaagsesse schunn uff em Disch. "Hock dich hie
un ess," saagt die Memm.

Des muss sie mir net zweemol heesse.

Vocabulary

sich aaduh, aageduh: to dress
aafange, aagfange: to begin
sich (nanner) aagreische, aagegrische: to yell at
   (each other)
aagucke: to look at
sich abdrickle, abgedrickelt: to dry oneself off
abkehre: to sweep off
sich an die Aerwet mache: to get to work, start in
   working
ausmischde: to clean a stable
ausschpanne: to unhitch a horse
sich balwiere, (ge)balwiert: to shave oneself
es Bee: leg
der Bese(m): broom
der Bese(m)schtiehl: broom handle
blaffich: said of dog in habit of barking
der Brei: porridge, pap
der Draemp, -s: tramp
der Droog, Dreeg: trough
sich dumm(e)le, gedummelt: to hurry
der Dummkopp, -kepp: stupid person
eischpanne: to hitch up a horse
feege: to sweep
es Gacki: child's word for egg
es Geilsgscharr (-gschaerr): harness
die Giesskann, -e: sprinkling can
heesse, gheesse: to request, bid, tell
sich hiehocke: to sit down
kehre: to sweep

time to get the harness and hitch up a horse to the
wagon. I have to take the ten o'clock lunch out to
Dad and my brothers. They must also have something to
drink and eat. Then I have to unhitch the horse again
and put the wagon back into the wagon shed.

Then I have to cut the grass in the yard, and the
morning is over. I go into the house and find the
lunch already on the table. "Sit down and eat," says
my Ma.

She doesn't have to bid me (do) that twice.

die Kett, -e: chain
der Kettehund, -e: dog kept on chain
 es Ketzli (Ketzel), -n: kitten
der Kiwwel: pail
 es Marriye-esse: breakfast
    melke, gemolke: to milk
der Melkschtuhl, -schtiehl: milking stool
der Mischt: manure
die Mischtgawwel, -gawwle: manure fork
der Mischthaahne: said of rooster free to roam barn-
       yard and manure pile
der Mischthaufe: manure pile
    mithelfe, mitgholfe: to help along
    nausfliege, nausgflogge: to fly out
    neischlaage, neigschlaage: to hit or knock in(to)
    rischde: to prepare (as food)
    ruffrufe: to call up to someone
    sich rumdrehe: to turn around
der Rumleefer: tramp
    schidde, gschitt: to pour
    schlaage, gschlaage: to hit
    schlabbe: to slop
der Schlabbkiwwel: slop pail
der Schloofkopp, -kepp: sleepy head
    schneide, gschnidde: to cut
    schtill: quiet, still
    sich schtreehle: to comb oneself
die Schulgleeder (pl.): school clothes
der Sunneschtraahl, -e: sun ray
    uffheere: to stop

      uffsei, uffgwest: to be up and about (out of bed)
      uffwecke: to wake up
      sich (druff) verlosse, verlosse: to depend (on
         something)
      sich verschloofe, verschloofe: to oversleep
      sich verschteck(e)le, verschteckelt: to hide one-
         self
      weckschtelle: to put away
die Weckuhr, -e: alarm clock
      weckwesche, weckgewesche: to wash away
      sich uff der Weg mache: to start off on one's way
 es Weggeli, -n: little wagon
      sich wesche, gewesche: to wash oneself
 es Zehuhrschtick: food eaten during a break in the
         morning's work. Also frequently referred to as
         es Neinuhrschtick.
      zufridde: satisfied

Grammar

A. Reflexive Verbs

1. In E, a reflexive verb is one that is followed
by such forms as myself, yourself, himself, etc. These
reflexive pronouns, as they are called, can be both
direct and indirect objects. For example, myself is a
direct object in the sentence I wash myself, but it is
an indirect object in the sentence I buy myself a hat.

2. In PG, reflexive verbs are also followed by a
specific set of reflexive pronouns, but those used as
indirect objects (dative case) differ in the 1st and
2nd persons singular from those used as direct objects
(accusative case). No dative reflexives were used in
the reading selections of this lesson; therefore, only
accusative forms will be taken up in the following discussion. The dative forms will be taken up in the next
lesson.

We normally recognize an E reflexive verb because
it is followed by "oneself," e.g., "to wash oneself."
We can recognize a PG reflexive verb because it is always preceded by <u>sich</u>, e.g., <u>sich wesche</u>.

3. The present tense indicative of <u>sich wesche</u> "to wash oneself," follows:

Singular	Plural
1. ich wesch mich	1. mir,mer wesche uns
2. du wescht dich	2. dihr,der wescht eich;
	ihr,er wesche eich; etc.
3. er,sie,es wescht sich	3. sie wesche sich

It is important to note that in the singular, <u>sich</u> is the third person reflexive pronoun for all three genders. Also, <u>eich</u> is the reflexive pronoun in the second person plural, no matter what form of the personal pronoun or verb the PG speaker uses.

4. When used with normal word order, the reflexive pronoun is always placed <u>right after</u> the verb. But a separable prefix or a past participle must still go to the end of the sentence (clause).

<u>sich rumdrehe</u> "to turn (oneself) around"
    ich dreh mich schnell rum
    du drehscht dich schnell rum
    etc.
    ich hab mich schnell rumgedreht
    du hoscht dich schnell rumgedreht
    etc.

5. When used with inverted word order, the reflexive follows the subject: <u>Nau wesch ich mich</u>. <u>Uffemol drehscht du dich rum</u>. <u>Dann hot er sich gewesche</u>.

By the way, perhaps you noticed that <u>hawwe</u> is the usual auxiliary for a reflexive verb; if a reflexive pronoun itself is not the direct object, then it will certainly be an indirect object, and another noun or pronoun will be the direct object, as we will see in our discussion of dative reflexive pronouns in the next lesson.

6. The two words <u>selwer</u> and <u>nanner</u> perform special functions when used with reflexive verbs and pronouns. The intensifier <u>selwer</u> is used for all persons, both singular and plural. It is thus the equivalent of the E intensifiers "myself," "yourself," "himself," "herself," "ourselves," "yourselves," and "themselves."

Note that in E the intensifiers are the same forms as the reflexives, and therefore they are very rarely if ever used together with reflexives. For instance, one would probably not hear anyone say "I will wash myself myself," or "I, myself, will wash myself." In E the intensifiers are more correctly used with objects other than reflexives: "I will wash the car myself," or "I, myself, will wash the car."

In PG, however, the intensifier <u>selwer</u> is used to reinforce the reflexive pronoun:

> Ich wesch mich selwer.
> Du wescht dich selwer.
> Mer wesche uns selwer, etc.

7. In PG if a reflexive verb is used to mean "each other," then <u>nanner</u> is added:

Sie gucke sich aa.	They look at themselves.
Sie gucke sich nanner aa.	They look at each other.
Sie finne sich.	They find themselves.
Sie hen sich nanner gefunne.	They found each other.

B. The Imperative Mood (Commands)

1. There is but one form of the imperative in English. It serves for both the singular and plural, and consists of nothing more than the verb. The subject of the command, "you," is always "understood"; that is, it is never actually spoken or written. We say, "Close the door, Mary," or "Jim and Joe, open the windows." The subject of "close" and "open" is <u>you</u>.

The PG speaker uses two forms of the imperative, a singular and a plural.

2.a. The singular form of the PG imperative, used when speaking to one person, is in effect the second person singular of the present tense without its ending -<u>scht</u>. For instance, the second person singular, present tense, form of <u>kumme</u> is <u>du kummscht</u>. The singular command is <u>kumm</u>. Likewise, the singular command derived from <u>du hockscht</u> is <u>hock</u>.

Just as in English, no pronouns are used with PG commands!

    b. However, if the stem of a verb ends in a sibilant (an s-like sound), then the singular command is nothing more than the verb stem: ess (from esse), wesch (from wesche), schwetz (from schwetze).

    c. The imperatives of verbs whose stems end in -d and -dd are usually written with -t and -tt, respectively, even though a rule of pronunciation states that a final d is pronounced like a t.

    d. The singular imperative of sei is always sei.

3.a. When speaking to two or more persons, the PG speaker uses plural commands that correspond to the various second person plural forms found on page 13. Thus, depending on where he lives, a PG speaker may say kumme, or kummt, or kummet. The suggestion in the paragraph following the conjugation of kumme on page is valid also in the case of imperatives.

    b. The plural imperative form of sei is always seid.

4. PG imperatives follow all rules concerning the placement of separable prefixes and reflexive verbs:

Schteh uff! (uffschteh)
Mischt die Schtell aus! (ausmischde)
Dummel dich! (sich dummle)
Hock dich hie! (sich hiehocke)
Drickelt eich schnell ab! (sich abdrickle)

CHAPTER TEN

Part Two

If you are not too tired from the full morning of
work, let's go on with the afternoon. Perhaps we'll
get a chance to rest a while, maybe even take a swim.
Nothing new in the way of grammar, but our vocabulary
will again contain a few animals and insects.

Die Sunn geht uff,
Kumm, Buwe, schteht uff,
Schunn lang sin die Haahne am Greehe;
Dihr faule Beng'le,
Macht eich ans Dengle,
Nooch'm Friehschtick geht's ans Meehe.

Harrich! 'S Middaag Hann geht,
Bis ans End watt's gemeeht,
Un dann geht alles nooch'm Haus,
'S Esse schmackt gut,
Ee kaze Schtunn watt's geruht,
Un noh uff's Feld widder naus.

Die Sunn iss ball nidder,
Die Nacht kummt ball widder,
Die Schadde warre lenger um em Bodde;
Die Kieh gehne heem,
Die Veggel nooch de Beem,
Aus de Lecher hupse die Grodde.

Gfiedert iss es Vieh,
Gemolke sin die Kieh,
Der Hund iss los vun der Kett;
Schliess die Dier zu,
'S iss Zeit fer nooch der Ruh,
Die Hinkel sin schun lang im Bett.[1]

[1] *Lee Light Grumbine, from "Der Alt Dengelschtock,"*
The Pennsylvania German Dialect.

En Gewehnlicher Daag; Es Zwett Deel

Darrich der Marriye waare die Weibsleit awwer aa
net faul. Sie hen beinaah all die Hausaerwet gemacht.
Es iss net Mundaag, awwer sie hen villeicht wennich gewesche. Un ich weess, sie hen gebiggelt; es Biggeleise
un Biggelbord sin noch in der Summerkich. Im Summer
iss es schee, in der Summerkich zu biggele. Der Keller
iss wennich dunkel.

Dann hen sie aa gebutzt; die Memm will immer alles
schee sauwer hawwe. Ich glaab, mer hen mehner Butzlumbe im Haus ass Weschlumbe--un mer hen abbaddich viel
vun denne! Un weil mei Schweschdere butze, do iss die
Memm am Backe. Sie will immer frieh mit ihrem Backes
faddich sei; dann kann sie es Middaag richde.

Nooch em Dinner neeht un flickt die Memm. Sie iss
ken Neehern, awwer kann so gut neehe ass (wie) eeni.
Un sie kann Hosse flicke un Schtrimp schtobbe besser
ass die menschde Weibsleit. Nau lannt sie mei Schweschdere schtrickle (schtricke).

Awwer mir Kinner kenne naus un schpiele. Mei gleener Bruder gleicht in der Gaarde zu geh un Keffer dot-
mache. Der Gremmpaepp hot em des gelannt; er selwer
kann sich nimmi gut bicke. Mei gleeni Schweschder dutt
des awwer gaar net gleiche. Sie watscht liewer die
Iemens (Uumens) wie sie schaffe, odder die Schpinne in
der Scheier. (Awwer mer farrichde (faerrichde) uns all
vor die Ieme un Weschbe; die kenne eem wiescht schteche,
un sell dutt verdollt weh.)

Mei Schweschder geht aa gaern in die Scheier un
schpielt mit der Katz un ihre Ketzlin. Unser Kaader
schpielt aa mannichmol mit. (Mer finnt ken Ratt odder
Maus in unsrer Scheier.) Un sie un der Billi ver-
schteck(e)le sich im Hoi un misse sich nanner finne.
Noh schpiele sie aa mit em Hutschel. Sie reide des
Geilche efders im Hof rum. Un efders falle sie runner,
duhne sich awwer nie net weh.

Alsemol gehne sie nunner zum Bechli, batsche im Wasser rum, un warre gans dreckich. Dann aergert sich
widder die Memm.

Ich kann aa unser Reitgaul unne Saddel reide. Ich
reit naus, wu der Paepp schafft, un dann losst er mich

A Typical Day; Part Two

But during the morning the women were also not lazy.
They did almost all of the housework. It isn't Monday,
but they probably washed a little. And I know they
ironed; the iron and the ironing board are still in
the summer kitchen. In summer it is nice to iron in
the summer kitchen. The cellar is a little dark.

Then they also cleaned; Mom always wants everything
nice and clean. I believe we have more dust cloths in
the house than wash cloths--and we have especially
many of them! And while my sisters clean, Mom is baking. She always wants to be finished with her baking
early; then she can prepare lunch.

After dinner (lunch) Mom sews and mends. She is no
seamstress, but can sew as well as one. And she can
mend pants and darn stockings better than most women.
Now she is teaching my sisters to knit.

But we children can go out and play. My little
brother likes to go into the garden and kill bugs. My
granddad taught him that, he himself can no longer
bend down well. My little sister, however, doesn't
like that at all. She rather watches the ants as they
work, or the spiders in the barn. (But we all fear
[are afraid of] the bees and wasps; they can sting you
awfully, and that hurts "a lot.")

My sister also likes to go into the barn and play
with the cat and her kittens. Our tomcat also plays
along sometimes. (One finds no rat or mouse in our
barn.) And she and Billy hide themselves in the hay
and must find each other. Then they play also with the
colt. They often ride that little horse around the
farm(yard). And often they fall off (down), but they
never hurt themselves.

Sometimes they go down to the brook(let), splash
around in the water, and get very dirty. Then my
mother gets angry again.

I can also ride my horse without a saddle. I ride
out where Dad is working, and then he lets me help

mithelfe. Odder ich reit nunner zum Deich un geh
schwimme un fische.

Die Zeit geht schnell, wann mer schpielt, un ball
sehn ich der Paepp un mei Brieder heemkumme. Des
meent, die Kinner misse aa uffheere schpiele. Es iss
ball Esszeit. Mer gehne all ins Haus; die Memm hot
schunn ihre Neehes--Neez, Noodel, un Scheer--weckgeduh,
un sie un mei Schweschdere hen schunn es Nachtesse
faddich.

Mer kumme runner mit sauwere Hend un Gsichder, un
do schtehne schunn die Deller mit Rinsfleesch un ge-
rooschdem Hinkel, un die Schissle mit Filsel, verdrick-
de (gemaeschde) Grummbeere, gedarrtem Welschkann, em
Dunkes (Greewie), un Grautsellaat.

Mer hen all en guder Abbeditt, un sell iss gut;
alles sehnt recht abbedittlich aus. Mer setze uns all
hie, awwer mer kenne noch net aafange. Aerscht misse
mer ruhich dositze; der Paepp iss schunn am Beede.

Er iss faddich. Un nau heert mer yuscht en Geglebb-
ber un en Gerabbel vun Gscharr. Un dann iss es widder
ruhich. Mer kann yuscht kaue heere. Kenner schwetzt.
Der Memm ihre Koches iss awwer ebbes Gudes.

Die Weibsleit nemme es Gscharr weck, awwer mer
schtehne noch net vum Disch uff. Nau kumme der Kuche
un der Boi uff der Disch. Un meh Millich un Kaffi.
In dem Haus verhungert nimmand. Do iss immer genunk
vun alles.

Awwer nooch em Owedesse kann mer sich net hielegge
un sich ausruhe. Mei eldschdi Schweschder wescht
oweds es Gscharr, un mei yunger Bruder un yungi
Schweschder misse abdrickle. Un noh iss es widder
Zeit, fer melke geh un fer die Hinkel un annere Diere
uff em Hof fied(e)re. Oweds geh ich liewer die Hinkel
fiedere. Im Hinkelhaus iss es immer heller ass wie in
der Scheier. Un ich heer nie ken Eil un sehn nie ken
Schpeckmaus driwwe am Hinkelhaus wie drunne an der
Scheier.

Es nemmt net lang, die Hinkel zu fiedre. Es watt
graad duschber; die Schwalme fliege hie un haer wie
der Blitz, un datt sehn ich schun en Feiervoggel. Un
ich geh graad ins Haus, do schtecht mich en Schnook.
Ich gratz mich gewiss die gans Nacht widder.

along. Or I ride down to the pond and go swimming and
fishing.

The time goes quickly when one plays, and soon I
see Dad and my brothers coming home. That means the
children also have to stop playing. It is soon time
to eat. We all go into the house; Mom has already put
away her sewing--thread, needle, and scissors--and she
and my sisters already have supper ready.

We come down with clean hands and faces, and there
already stand the plates with beef and roast chicken,
and the bowls with filling, mashed potatoes, dried
corn, the gravy (Dunkes from dunke "to dunk"), and
cole slaw.

We all have a good appetite, and that is good;
everything looks right appetizing. We all sit down,
but we can't start yet. First we must sit here quietly; Pop is already praying.

He is finished. And now one hears just a clatter
and rattle of dishes. And then it is quiet again.
One can just hear chewing. Nobody talks. Mom's cooking is something good.

The women take away the dishes, but we don't get up
from the table yet. Now come the cake and the pie on
the table (are put on the table). And more milk and
coffee. In this house no one starves. There is always enough of everything.

But after supper one can not lie down and rest. My
oldest sister washes the dishes in the evening and my
young brother and young sister must dry. And then it
is again time to go milking and to feed the chickens
and other animals on the farm. In the evening I
rather go feed the chickens. In the chicken house it
is always lighter than in the barn. And I never hear
an owl and see a bat over at the chicken house like
down at the barn.

It doesn't take long to feed the chickens. It is
just getting dusk; the swallows fly back and forth
like lightning, and there I already see a lightning
bug. And I am just going into the house, a mosquito
stings me. I'll certainly scratch myself the whole
night (all night) again.

Es iss ball Bettzeit. Ich geh nuff un geh in mei
Schloofschtubb nei. Ich kann die Memm singe heere;
sie singt em Bobbli en Schockellied. Nau zieh ich
mich aus, muss awwer noch wennich waarde. Mei
Schweschder iss noch die zwee Gleene am Baade. Ball
sin sie awwer haus, un ich kann nei. Nadierlich kann
ich mich selwer baade.

Des nemmt aa net lang, kann ich der saage (ich hab
yo marriye ken Schul), un noh kann ich noch mit mei
Gschwischder un der Wuffi schpiele. Awwer mer sin all
mied un schleefrich. Mei Bruder un ich saage "Gude
Nacht," gewwe em Paepp un der Memm en Boss, un gehne
nuff ins Bett.

Was? Wu iss der Wuffi? Ya, kannscht des net rode?
(Ruhich, Wuffi. Sei schtill, un muck dich net!)

Vocabulary

```
    abbedittlich:  appetizing
    sich ausruhe:  to rest
    sich ausziehe, ausgzogge:  to undress
    baade:  to bathe
    sich baade:  to bathe oneself
    beede, gebeedt:  to pray
    sich bicke:  to bend down
 es Biggelbord:  ironing board
    biggele:  to iron
 es Biggeleise:  flatiron
der Boss, -e:  kiss
der Butzlumbe:  dust rag, cleaning rag
    dositze, dogsesse:  to sit here
    dotmache:  to kill
    dreckich:  dirty
 es Dunkes:  gravy
die Eil, -e:  owl
    sich farrichde (faerrichde):  to be afraid
    faul:  lazy
der Feiervoggel, -veggel:  lightning bug
    fied(e)re:  to feed (animals)
 es Fillsel:  filling
    sich (nanner) finne:  to find (each other)
    flicke:  to mend
    gedarrt (gedaerrt):  dried
```

It is soon bedtime. I go up and go into my bedroom. I can hear Mom singing; she is singing the baby
a lullaby. Now I get undressed, but must wait a
little. My sister is still bathing the two little
ones. But soon they are out, and I can (go) in.
Naturally I can bathe myself.

That doesn't take long, I can tell you (after all,
I don't have school tomorrow), and then I can play
with my siblings and Wuffi. But we are all tired and
sleepy. My brother and I say "Good night," give my
Dad and Mom a kiss, and go up to bed.

What? Where is Wuffi? Well, can't you guess that?
(Quiet, Wuffi. Be still, and don't move!)

es Geglebber: clattering
es Geilche: little horse, pony
  gemaescht: mashed
es Gerabbel: rattling
  gerooscht: roasted
  sich gratze: to scratch oneself
es Grautsellaat: cole slaw
es Greewie: gravy
  Gude Nacht: good night
die Hausaerwet: housework
  sich hieleege, hiegeleegt: to lie down
  sich hiesetze: to sit down
  hie un haer: back and forth
die Hosse (pl.): pants
die Iem, -e: bee
die Iemens (Umens), -e: ant
der Kaader: tomcat
  kaue: to chew
der Keffer: bug
  mitschpiele: to play with, to play along
  sich mucke: to move (oneself)
die Neehern: seamstress
es Neehes: sewing
der Neez: thread
die Noodel, Noodle: needle
die Ratt, -e: rat

reide, geridde: to ride
es Rinsfleesch (Rindfleesch): beef
rode: to guess
ruhich: quiet(ly)
rumbatsche: to splash around (in water)
runnerfalle: to fall down (from)
der Saddel, Seddel: saddle
die Scheer, -e: scissors
schleefrich: sleepy
die Schnook, Schnooge: mosquito
es Schockellied, -lieder: lullaby
die Schpeckmaus, -meis: bat
die Schpinn, -e: spider
schteche, gschtoche: to sting; to stick, prick
schtobbe: to darn
schtrickle, gschtrickelt (schtricke, gschtrickt):
    knit
der Schtrump, Schtrimp: stocking
die Schwalm, -e: swallow
verdollt: confounded(ly)
verdrickt: mashed, squashed
verhungere: to starve
watsche: to watch
weckduh, weckgeduh: to put away
weh: hurt
sich weh duh: to hurt oneself
die Weschp, Weschbe: wasp

CHAPTER ELEVEN

Part One

Much of the food found on the table of the PG farm
family is raised right on the farm, but occasionally a
trip to town is necessary when something like coffee or
sugar runs low. Our young friend is growing up; it is
time to entrust him with a trip into town.

And it is time for us to take the giant step to
transposed word order, found in all PG subordinate
clauses beginning with either subordinating conjunctions or relative pronouns. The finite verb forms that
appear at the very end of clauses or sentences indicate
the transposed word order of the subordinate clause.

*Am Greizweg schteht's Wattshaus un der Schtor
in eem lange Gebei, un en Bortsch drei Dreppe
hoch mit Dach druff am ganse Gebei draa naus.
Vor der Barschtubb Dier sin en paar alde Schtiehl
un en langi Bank; do sitze die "Rumhenker" un die
Bauere im Summer, wann sie beikumme, ihre Ge-
schefte zu mache, fer en bissel blaudre un fer
Neiichkeede ausfinne un bringe.*

*Am Eck vum Wattshaus schteht der Schildposchde
un owwedruff iss en Schild mit me groosse schwaze
beessguckiche Baer, uff beede Seide gemoolt. Er
schteht uff de hinnere Fiess mit em Rache grooss
uff.*

*Uff em annere Eck an de seeme Seit vun der
Schtrooss iss der Schmittschapp. En Weil zrick
hot der Schmitt drei Feiere am Geh ghatt, un noch
en Brendis debei, awwer zidder ass die Bauere uff
em Feld rumyaage mit Draeckders, nemmt es nimmi
so viel Hufeise, un der Schmitt schafft alles.*

(Continued, Part Two)

Mir Gehne Ins Schteddel; Es Aerscht Deel

Ich muss der mol verzeehle, wie mei gleener Bruder
un ich geschder ins Schteddel neigefaahre sin. Nadierlich waar ich schunn efders im Schteddel gwest, awwer
es iss net oft (ge)bassiert, ass ich allee ins Schteddel gange bin. Awwer ich wa nau elder, alt genunk, so
ass ich selwer mit meim Beiseckel odder Weggeli neifaahre kann. (Deel Leit wisse, ass Faahrraad es
deitsch Watt fer Beiseckel iss.)

Mer waxe fascht alles, was mer esse, awwer alsemol
hot die Memm schnell Zucker (Sucker) odder Tee, odder
so ebbes hawwe misse. Dann hot sie als waarde misse,
bis der Paepp heemkumme iss, un er hot dann mit der
Maschien ins Schteddel faahre misse.

Eb mer uns uff der Weg noch em Schteddel gemacht hen,
hen mei Bruder un ich uns aerscht wesche un schtreehle
misse. Nadierlich hen mer gemault. Awwer weil die
Memm der Billi gewesche un gschtreehlt hot, hab ich mir
selwer 's Gsicht gewesche un mir 's Haar geschtreehlt.
Der Billi hot als noch gemault, awwer nix, was er ge-
saat hot, hot gebatt. "Halt schtill! Maul net! Ich
will net, ass dihr im Schteddel rumlaaft, wie en paar
Rumleefer. Dihr wisst yo, was sie vun der Williamsen
saage."

Ya, die Memm hot recht ghatt. Des iss die Fraa,
daere ihre Buwe un Meed immer so schlabbich un dreckich
im Schteddel rumlaafe. Villeicht sin sie Englisch.

Bis die Memm mit em Billi faddich waar, waar ich
schunn draus un hab der Gaul eigschpannt ghatt. Mer
sin mit em Weggeli gfaahre, weil der Billi noch zu yung
iss, mit em Beik (Raad) ins Schteddel zu faahre. Wann
er rauskumme iss, iss er glei eigschtigge; un noochdem
mir der Memm noch en Kuss gewwe hen ghatt, sin mer abgfaahre. Wie mer uff die Schtrooss gfaahre sin, hen
mir die Memm rufe heere, "Basst uff! Un faahrt net so
schnell!" Awwer ich bin schee vorsichtich gfaahre, un
der Billi hot sich gaar net gfarricht (gfaerricht).

Mer hen net lang faahre misse, do hen mer schunn der
Karrichetarn gsehne. Es Schteddel hot ken groosse
Affissgebeier, wie die, ass in der Schtadt sin. Die
menscht sin zwee- odder dreischteckich, un der Tarn iss
viel heecher. Nau waare mer schunn an der alde Miehl.

We Go Into Town; Part One

I must tell you how my little brother and I drove
into town yesterday. Naturally I had already been in
town often, but it didn't happen often that I went into
town alone. But I'm getting older now, old enough so
that I can drive in myself with my bicycle or wagon.
(Some people know that <u>Faahrraad</u> is the German word for
bicycle.)

We grow almost everything that we eat, but sometimes
Ma had to have sugar or tea, or something like that
quickly. Then she used to have to wait till Dad came
home, and he then had to drive into town with the car.

Before we started on our way into town, my brother
and I had to wash ourselves and comb ourselves first.
Naturally we grumbled. But while Ma washed and combed
Billy, I washed my face and combed my hair. Billy was
still grumbling, but nothing that he said helped.
"Hold still! Don't grumble! I don't want that you
walk around in town like a couple of tramps. You know
what they say about the Williams woman."

Yes, Ma was right. That is the woman whose boys and
girls always walk around in town so sloppy and dirty.
Maybe they are "English." (See vocabulary note.)

Till Ma was finished with Billy, I was already outside and had hitched up the horse. We drove with the
wagon, because Billy is still too young to go (drive)
into town on his bike. When he came out, he climbed in
right away; and after we had given Ma a kiss, we drove
off. As we drove onto the street, we heard Ma calling,
"Watch out (be careful)! And don't drive so fast!"
But I drove nice and carefully, and Billy didn't get
afraid at all.

We didn't have to drive long, when we already saw
the church tower. The town has no large office buildings, like those that are in the city. The most are
two- and three-storied, and the steeple is much higher.
Now we were already at the old mill. It is closed, and

Sie iss zu, un es Miehlraad geht nimmi rum. Un mer
kann neddemol sehne, wu es Miehlrees waar. Mer hen
mol gheert, ass groosse Katzefisch im Miehldeich sin.
Do hen mer fische geh wolle, awwer die Kinner, ass datt
wuhne, un denne ihre Vadder der Miller waar, hen uns
fattgeyaagt. Sie hen des geduh, so ass mer kenni vun
ihre Katzefisch fange hen kenne. Awwer die Katzefisch,
ass do drin im Deich sin, sin ennihau zu glee.

Wie mer ins Schteddel kumme sin, hen mer unser Gaul
an en alder Geilsposchde aagebunne. Der Aabinnposchde
schteht wu en Wattshaus mol gschtanne hot ghatt. Es
Wattshaus (Waertshaus) hot en Barschtubb ghatt, un aa
en groossi Dansschtubb, wu die Leit, ass do rum wuhne,
Freidaag un Samschdaag oweds gedanst hen. Un do waar
als en groossi Schild uff me Schildposchde, ass aa do
vanne gschtanne hot, un es hot en Barr (Baer) druffgemoolt ghatt. Ich weess nimmi, eb der Barr blo odder
schwaz waar. Ich muss mol mei Paepp frooge.

Awwer es Wattshaus iss en paar Yaahr zrick gans nunnergebrennt. Ich kann mich noch errinnere (aerrinnere)
wie die Flamme hoch in die Luft gschosse sin. Mer hen
sie gut sehne kenne, wann mer aus em Fenschder owwe im
Schpeicher geguckt hen. Zidder ass des bassiert iss,
sin der Watt un sei Familye in die Schtadt gezogge.
Fascht yedi Fraa, daere ihre Mann schtunnelang im Wattshaus gsotze iss, hot en net gegliche, awwer die Mannsleit hen immer gsaat, ass vun all denne Wadde im Kaundi,
waar aer der bescht.

Vocabulary

    aabinne: to tie up to
der Aabinnposchde: hitching post
    abfaahre, abgfaahre: to drive off
 es Affissgebei, -er: office building
    allee: alone
    badde, gebatt: to help, avail
der Barr (Baer), -e: bear
 es Beik, -s: bike
 es Beiseckel, -s: bicycle
die Dansschtubb, -schtuwwe: dance hall
    druffmoole: to draw, paint on
    eischteige, eigschtigge: to climb into a vehicle

the mill-wheel no longer goes 'round. And one can't even see where the mill-race was. We once heard that large catfish are in the mill pond. We wanted to go fishing, but the children who live there, and whose father was the miller, chased us away. They did that so that we could not catch any of their catfish. But the catfish that are in the pond are too small anyhow.

When we got into the town, we tied our horse to an old hitching post. The hitching post stands where an inn once had stood. The inn had a barroom, and also a large dance hall where the people who lived around here danced on Friday and Saturday evenings. And there used to be a big sign on a signpost, that also stood in front here, and it had a bear drawn (painted) on it. I don't know anymore whether the bear was blue or black. I must ask my Dad some time.

But the inn burned down completely a few years ago. I can still remember how the flames shot high into the air. We could see them well when we looked out of the window upstairs on the second floor. Since that happened, the innkeeper and his family have moved into the city. Almost every wife whose husband sat for hours in the inn did not like him, but the men always said that of all the innkeepers in the county, he was the best.

     Englisch: anyone not PG (see note)
     sich errinnere (aerrinnere): to remember
 es Faahrraad, -redder: bicycle
     fattyaage: to chase away
     fische: to fish
der Geilsposchde: hitching post
der Karrichetarn, -e: church steeple, tower
der Katzefisch: catfish
der Kuss, Kiss: kiss
die Maschien, -s: car
     maule: to grumble, "mouth off"
die Miehl, -e: mill
der Miehldeich, -er: mill pond
 es Miehlrees: mill race

der Miller: miller
   neifaahre, neigfaahre: to drive in
 es Raad, Redder: wheel, bicycle
   rumgeh, rumgange: to go around
 es Schild, -er: sign
der Schildposchde: signpost
   schlabbich: sloppy, sloppily
   -schteckich: storied; zweeschteckich: two-
     storied, two floors high
   sitze, gsotze: to sit
der Tarn, -e: tower, steeple
   verzeehle: to relate, tell
   vorsichtich: careful(ly)
 es Wattshouse (Waertshaus), -heiser: inn
der Weg, Weege: road, way, path
   sich uff der Weg mache: to start off

Vocabulary Notes

die Maschien: also die Kar, Keer, Aademobiel, Addeme-
bill.

der Tarn: also der Tarm, der Turm.

Englisch: Pennsylvania Germans refer to people who
are not PG as English, no matter what their true
nationality. This idea is not without precedent;
we are reminded that historians tell that so many
of the Germans who arrived in Philadelphia after
1730 were from the Palatinate that all Germans were
referred to as the Palatines.

Grammar

A. Reflexive Verbs; Dative Reflexive Pronouns

In Chapter Ten we discussed the use of reflexive
verbs whose accompanying reflexive pronouns are direct
objects and therefore in the accusative case.

In the sentence Ich wesch mir die Hend "I wash my
hands (I wash for myself the hands), Hend is the direct
object of the verb wesch, and the reflexive pronoun mir
is thus the indirect object. You can be rather certain

that if a reflexive verb has an object <u>other than</u> the reflexive pronoun, then the reflexive pronoun will actually be an indirect object (dative case).

1. Dative reflexive pronouns used as indirect objects differ from accusative objects in the 1st and 2nd persons singular only, as the following conjugation of <u>sich die Hend wesche</u> "to wash one's hands" shows:

Singular	Plural
1. ich wesch mir die Hend	1. mir,mer wesche uns die Hend
2. du wescht dir die Hend	2. dihr,der wescht eich die Hend; ihr,er wesche eich die Hend, etc.
3. er,sie,es wescht sich die Hend	3. sie wesche sich die Hend

Once again, note that <u>sich</u> is the reflexive form for all three genders in the third person singular, and that <u>eich</u> is the reflexive for all personal pronouns in the second person plural.

2. The dative forms must also be used when the verb takes dative objects, as for instance <u>helfe</u> "to help": <u>ich helf mir,mer (selwer)</u>, <u>du helfscht dir,der (selwer)</u> "I help myself, you help yourself."

B. Subordinate Clauses

In PG, as indeed in E, subordinate clauses (also called dependent clauses) may begin with subordinating conjunctions or with relative pronouns. The subordinating conjunctions and their influence on the word order of the clauses they introduce will be discussed in this section of Chapter Eleven. Relative pronouns will be taken up in Part Two of this chapter.

1. Subordinating Conjunctions

a. The following subordinating conjunctions are found in PG. The list is merely alphabetical, thus those that are at the head are not necessarily more common or important than are those that follow.

```
ass:  that
bis:  till, until; by the time (that)
eb:  before; whether
noochdem:  after
so ass:  so that, in order that
wann:  when; if
weil:  while; because
wie:  as, when, how
wu:  where
zidder ass:  since
```

Two more such subordinating conjunctions--<u>bevor</u> "before" and <u>soball</u> "as soon as"--are not very common, but are occasionally found in PG literature.

In addition, <u>ass (wie) wann</u> "as if" is used with the subjunctive mood (contrary to fact conditions) and will be taken up in conjunction with the subjunctive mood in a later lesson.

b. Let us now look at these conjunctions and their uses as illustrated in clauses from the preceding reading selection.

<u>ass</u> "that" (also <u>dass</u>, especially in literature)

Es iss net oft bassiert, ass ich allee ins Schteddel bin.	It didn't happen often that I went into town alone.

<u>bis</u> "until," "till"

Dann hot sie als waarde misse, bis der Paepp heemkumme iss.	Then she used to have to wait till Dad came home.

<u>bis</u> "by the time"

Bis die Memm mit em Billi faddich waar, waar ich schunn draus.	By the time Mom was finished with Billy, I was already outside.

<u>eb</u> "before"

Eb mer uns uff der Weg gemacht hen, hen mei Bruder un ich uns wesche misse.	Before we started on our way, my brother and I had to wash ourselves.

eb "whether"

Ich weess nimmi, eb der Barr schwaz waar.	I don't know any more whether the bear was black.

noochdem "after"

Noochdem mir die Memm en Boss gewwe hen ghatt, sin mer abgfaahre.	After we had given Mom a kiss, we drove off.

so ass "so that"

Ich bin alt genunk, so ass ich selwer ins Schteddel faahre kann.	I am old enough so that I can drive into town alone.

wann "when," "if"

Wann er rauskumme iss, iss er glei eigschtigge.	When he came out, he immediately climbed in.

weil "because," "while"

Mer sin mit em Weggeli gfaahre, weil der Billi noch zu yung iss.	We drove with the little wagon because Billy is still too young.

wie "how," "when," "as"

Ich kann mich noch errinere, wie die Flamme in die Luft gschoosse sin.	I can still remember how the flames shot into the air.

wu "where"

Mer kann net sehne, wu es Miehlraad waar.	One cannot see where the mill wheel was.

zidder ass "since"

Zidder ass des bassiert iss, iss der Watt in die Schtadt gezogge.	Since that happened, the innkeeper has moved into the city.

In addition, all interrogatives (question words) can be used as subordinating conjunctions: Wieviel (wiffel) Leit sin do? Ich weess net, wieviel Leit do sin.

C. Transposed Word Order in Subordinate Clauses

1. Many of the sentences used in B can again be used
to illustrate the construction of subordinate clauses.

You have no doubt noticed that throughout the first
reading selection the finite verb* is placed last in a
subordinate clause:

| Bis die Memm mit em | By the time Mom was |
| Billi faddich waar, ... | finished with Billy, ... |

However, the truth of the matter is that PG speakers
many times disregard such formal "rules" and frequently
place a modifier such as a prepositional phrase last:

Bis die Memm faddich waar mit em Billi, ...

2. Compound Verbs in Subordinate Clauses: When used
to form the present perfect tense, the auxiliaries <u>sei</u>
and <u>hawwe</u> are placed at the end of subordinate clauses,
immediately following the past participle of the main
verb:

Wann er rauskumme iss,..	When he came out, ...
Eb mer uns uff der Weg	Before we set out on our
gemacht hen, ...	way, ...

But when used to form the past perfect tense, these
auxiliaries are placed <u>between</u> the two past participles:

| Noochdem mir der Memm en | After we had given Mom a |
| Boss gewwe hen ghatt,... | kiss, ... |

3. A modal auxiliary accompanied by a dependent infinitive follows the rules as outlined above:

..., so ass ich selwer	..., so that I can drive
ins Schteddel faahre	into town myself.
kann.	

---

*
 A finite verb is a verb form that changes from person
to person and from number to number. For instance, in
the following compound verbs, the auxiliaries change
according to number and person, whereas the past participles never change: <u>ich bin gange, du bischt gange,
mer sin gange; ich hab kaaft, du hoscht kaaft, mer hen
kaaft</u>. Both past participle and infinitives, verb
forms that never change, are called verbals.

But a modal auxiliary and a dependent infinitive used in the present perfect tense are always "split" by the auxiliary <u>hawwe</u> (the finite verb in this construction):

Sie hen sell geduh, so ass mer kenni vun ihre Katzefisch fange hen kenne.	They did that so that we could not catch any of their catfish.

4. When a main clause (also called an independent clause) precedes a subordinate clause, it follows all the rules of normal or inverted word order that have already been discussed in previous chapters. However, if a main clause follows a subordinate clause, then it regularly has <u>inverted</u> word order, whether or not an element other than the subject comes first:

Wann er rauskumme iss, iss er glei eigschtigge.	When he came out, he immediately climbed in.

5. And lastly, notice that in written PG, subordinate clauses are set off from main clauses with a comma:

Mer kann net sehne, wu es Miehlraad waar.	One cannot see where the mill-wheel was.

CHAPTER ELEVEN

Part Two

Now that we are familiar with the oddity of a finite verb at the <u>end</u> of a clause, we should have no trouble with word order in this second section of our lesson. Relative pronouns are rather easy to learn; there are not that many of them. And, wie gsaat, we are already familiar with the word order of the relative clause, the second type of subordinate clause in the PG dialect.

Graad iwwers Eck vum Wattshaus iss die Kaerrich, en groossi, aldi, schteeni gemeescheftlichi Kaerrich mit me Tarn, net iwweraus hoch, owwe druff, ass sie so bissel gucke macht wie en aldi, dicki Mudder, mit me gleene Hietli uff. Der Begreebnisblatz iss um die Kaerrich rum bis hinnenaus an der Busch, so wie die gleene Kinner un Kinskinner sich um die Mammi neschtle.

Uff em vierde Eck vum Greizweg iss en Dokder sei Haus. Des iss der reich Mann vun der Nochberschaft. Der Graabschteehacker wuhnt mit seim Schapp newedraa, en glee Schtick uff der annere Seit der Kaerrich.

'S Wattshaus, der Schtor un die Kaerrich sin, so wie mer saagt, yuscht ebaut die Mitt vun der Welt, fer die Leit, wu etliche Meil do rum wuhne.[1]

---

[1] Lloyd A. Moll, from <u>Am Schwarze Baer</u>.

Mir Gehne Ins Schteddel; Es Zwett Deel

Wie gsaat, es Schteddel iss gaar net grooss. Am
Greizweg schteht en Schtor, die Karrich, em Dochder
sei Haus, un en Gebei, wu der Balwierer un der Schuhmacher (Schuschder) ihre Schapps hen. Mer hen aa en
Feierhaus, wu die schee, rot Feierinschein iss. Awwer
zidder ass es Wattshaus nunnergebrennt iss, denke deel
Leit, ass sie sich net uff die Feierkump(e)ni verlosse
kenne.

Un do waar mol en Muviehaus do im Schteddel, awwer
sie iss aa zu, un mer muss nau in die Schtadt faahre,
wann mer noch de Muvies geh will. Awwer die Leit, ass
Muvies gleiche, kenne sie yo heidesdaags uff em Guckbox sehne.

Unser Karrich iss arrig alt. Mei gansi Familye un
viel vun de Leit, ass unsere Freinde sin, sin Gemeens-
glieder vun daere Karrich. Mer hen graad letscht Yaahr
en neier Parre grickt. Der alt Parre hot der Gemee
lang gedient. Mei Eldre hen en arrig gegliche; er hot
sie yo gheiert. Awwer ich gleich der nei Parre. Er
breddicht net so lang, un weil sei Breddiche graad halb
so lang sin, singe mer mehner vun denne scheene alde
Lieder.

Newwich der Karrich iss der Karrichhof, wu mei
Groossgraenpaepp (Urgroossvadder) un Groossgraenmemm
(Urgroossmudder) begraawe sin. Ihre Greewer sin graad
uff der anner Seit der Mauer. Mer kann ihre Naame, ass
uff em Graabschtee sin, gut lese, awwer deel Graabschtee sin so alt, ass mer die Naame, ass druff sin,
beinaah net lese kann.

Noochdem mer unser Gaul aagebunne hen ghatt, simmer
in der Greizwegschtor neigange. Der Mann, ass der
Schtorkipper vum Schtor iss, iss aa unser Poschtmeeschder. Sei Affiss iss graad im Schtor drin. Der Billi
iss gleich hiegange, wu es Kaendi waar. Awwer ich hab
aerscht schnell Kaffi un Zucker kaafe misse. Wie ich
niwwer zum Billi gange bin, waar er immer noch am Gucke.
Er hot es Kaendi net finne kenne, ass er geguckt hot
defor, un hot beinaah aagfange zu heile. (Der Billi
saagt immer, waer ken Kaendi gleicht, daer wees net, was
gut iss.) Awwer wann der Billi Geld hot, ass er schpende kann, dann muss er's schpende. Noh hot er sich en

We Go Into Town; Part Two

As I've said, the town is not big at all. At the crossroad stands a store, the church, the doctor's house, and a building where the barber and the shoemaker have their shops. We also have a firehouse where the beautiful red fire engine is. But since the inn burned down, the people think that they can not rely (depend) on the fire company.

And there was once a moviehouse here in town, but it is also closed, and one must now drive into the city if one wants to go to the movies. But the people who like movies can of course see them these days on television.

Our church is very old. My whole family and many of the people that are our friends are church members of this church. We just got a new preacher last year. The old preacher served the congregation for a long time. My parents liked him a lot; he married them. But I like the new preacher. He doesn't preach as long, and because his sermons are just half as long, we sing more of those beautiful old songs.

Beside the church is the cemetery where my great-grandfather and greatgrandmother are buried. Their graves are just on the other side of the wall. One can read well their names that are on the gravestone, but some gravestones are so old that one almost can't read the names that are on them.

After we had tied up our horse, we went into the crossroads store. The man who is the storekeeper of the store is also our postmaster. His office is right in the store. Billy went right over (to) where the candy was. But I first had to buy coffee and sugar. When I went over to Billy, he was still looking. He could not find the candy that he was looking for, and almost began to cry. (Billy always says whoever doesn't like candy doesn't know what's good.) But when Billy has money that he can spend, then he must spend it.

paar Zuckerschtengel rausgsucht (die hen yuscht zwansich Zent koscht). Ich hab mir aa ebbes Siesses rausgsucht, un hab dann fer alles, was mer kaaft hen, bezaahlt. Wann mer noch em Schtor gehne, fer Lewesmiddel
odder ebbes schunnscht zu kaafe, bezaahle mer immer
Kaesch defor.

Der Schtorkipper hot alles in en Babiersack, en
Dutt, geschteckt, un noh sin mer naus zum Weggeli, hen
der Gaul vum Aabinnposchde losgemacht, sin ins Weggeli
eigschtigge, un hen uns uff der Heemweg gemacht.

Alles iss gut gange, mer hen ken Unglick (Aeksident)
ghatt, un mer sin gut heemkumme. Awwer die wieschde
Kinner an der Miehl hen uns en Schtee noochgschmisse,
un was ich ne noochgerufe hab, kann ich der gaar net
saage. Ich hab em Billi schunn gsaat, ass er der Memm
nie verzeehle soll, was ich gsaat hab. Was sie net
weess, macht ihr net heess.

Wie mer widder deheem waare, waar die Memm froh, uns
widder zu sehne. Un ich weess, ass sie uns ball widder
geh losst.

Vocabulary

  es Affiss: office
der Balwierer: barber
    begraawe, begraawe: to bury
  es Breddich, -e: sermon
    breddiche: to preach
    diene (+ dative object): to serve
  es Feierhaus, -heiser: firehouse
die Feierinschein, -e: fire engine
die Feierkump(e)ni, -s: fire company
die Gemee: congregation
  es Gemeensglied, -er: member of a congregation
  es Graab, Greewer: grave
der Graabschtee: gravestone
der Greizweg, -e: crossroads
der Greizwegschtor, -s: store at a crossroad
  es Guckbox, -e: television set
    heile: to cry
  es Kaesch: cash
    koschde, koschdt: to cost

Then he picked himself out a few sticks of candy (they cost only twenty cents). I also picked out something sweet, and then paid for everything that we bought. When we go to the store to buy food or something else, we always pay cash for it.

The storekeeper put (stuck) everything into a paper sack, a bag, and then we went out to the wagon, untied the horse from the hitching post, climbed into the wagon, and started off for home.

Everything went well, we didn't have any accident, and we got home o.k. But those bad children at the mill threw a rock after us, and what I called after them I can't tell you at all. I have already told Billy that he should never tell Mom what I said. What she doesn't know won't hurt her.

When we were home again, Mom was happy to see us again. And I know that she will soon let us go again.

  es Lied, Lieder: song
     losmache: to untie, to loosen
  es Muviehaus, -heiser: movie house
 die Muvies: movies
     noochrufe, noochgerufe: to call after
     noochschmeisse, noochgschmisse: to throw after
 der Parre: preacher, parson
 der Poschtmeeschder: postmaster
     sich raussuche: to pick out (for oneself)
 der Schapp, -s: shop
     schpende, gschpendt: to spend
 der Schtorkipper: storekeeper
 der Schuhmacher: shoemaker
     schunnscht: otherwise; ebbes schunnscht: some-
        thing else
 der Schuschder: shoemaker
  es Siesses: sweets
  es Unglick, -e: accident
 der Zuckerschtengel, -le: candy stick

224

Vocabulary Notes

der Karrichhof: also der Graabhof, der Begreebniss-
    blatz.

der Parre: also der Paarre, der Porre.

Grammar

A. Relative Pronouns

1. In E, relative pronouns are who (whose, whom),
that, and which: He is the man who saw us today; he is
the man whom we saw today; he is the man whose wife we
saw today; this is the book that we must read; the book
which we must read is interesting.

2. The equivalent PG pronouns are either ass or wu.
In "The Pennsylvania German Dialect," which appears as
an appendix to The Pennsylvania Germans, edited by
Ralph Wood, Albert F. Buffington writes that a morpho-
logical peculiarity which he regards as distinctively
Pennsylvania German is "the extensive use of ass as a
relative pronoun in all cases of the singular and plural." And in A Pennsylvania German Grammar, coauthored
by Buffington and Preston A. Barba, Buffington writes:
"A distinctive feature of the Pennsylvania German dialect spoken today in the various sections of Pennsylvania is the extensive and increasing use, particularly
by the younger generations, of ass as a relative pronoun for all genders and the common cases [nominative
and accusative cases] of the singular and plural . . ."

But be that as it may, the fact remains, however,
that the relative pronoun wu is used extensively by
many PG writers. (For example, in Lloyd A. Moll's Am
Schwarze Baer one can find wu used as a relative pronoun five times in four successive sentences.) Thus
for the benefit of those who go on to the reading of PG
literature, the following examples will include wu, although this relative has not been used in the reading
selections. (The relative pronoun wu is not to be confused with the subordinating conjunction or interrogative wu "where.") Note the use of transposed word
order.

Der Mann, ass (wu) der Schtorkipper iss ...	The man who is the store-keeper ...
Die Fraa, ass (wu) mer gsehne hen ...	The woman whom we saw ...
Die Leit, ass (wu) Mu-vies gleiche ...	The people who like movies ...

3. We have already noted that the genitive case is missing in PG, and that a dative case form followed by a possessive pronoun takes its place (see page 40). The PG equivalent of the E possessive relative pronoun "whose" is constructed in much the same way, except that the dative case of the demonstrative is used. The gender and number of the demonstrative must agree with the antecedent, and it must once again be followed by a possessive that also agrees with the number and gender of the antecedent:

Des iss die Fraa, daere ihre Kinner ...	That is the woman whose children ...
Der Mann, dem sei Fraa ...	The man whose wife ...
Die Kinner, denne ihre Vadder ...	The children whose father ...

4. The PG equivalent of "he who" or "whoever" is the indefinite relative pronoun waer:

Waer ken Kaendi gleicht, (daer) weess net, was gut iss.	Whoever (he who) doesn't like candy doesn't know what's good.

The demonstrative daer beginning the main clause has been placed in parentheses to indicate that it is not absolutely necessary; its function is to emphasize.

But if the verb of the main clause takes a dative object, then the demonstrative must be used:

Waer des Kaendi gleicht, dem gebbich en Schtick.	Whoever likes this candy, to him I'll give a piece.

As happens so often, a literal translation is, of course, very awkward; it would best be rendered as "I'll give a piece of candy to anyone who likes it."

5. Like the interrogative pronoun waer, the relative pronoun waer has a dative form wem. Occasionally one

will hear--and, especially, read--an accusative form
wen.

6.a. The PG equivalent of E "what" or "that which"
is the indefinite relative pronoun was:

Was nei iss, iss net immer besser.	What is new is not always better.

b. Was is always neuter and is both the nominative
and accusative form.

c. Was is used also as the relative pronoun when
the antecedent is an indefinite pronoun such as alles
or nix:

Alles, was er gsaat hot ...	Everything that he said ...
Nix, was er gsaat hot...	Nothing that he said ...

7. In PG it is not usual for a relative pronoun to
be the object of a preposition. Rather, de- is placed
before the preposition involved, and the resulting compound is placed at the end of the relative clause:

Er hot es Kaendi net finne kenne, ass er geguckt hot defor.	He could not find the candy for which he was looking (that he was looking for).

Subordinate clauses beginning with relative pronouns
follow all the rules of word order discussed under the
heading Word Order in Subordinate Clauses.

CHAPTER TWELVE

Part One

We are rapidly coming to the end of the list of grammatical concepts necessary for a complete understanding of PG. Two of these final concepts are the future and the future perfect tenses. The text takes up some of the vocabulary of the city, filled, as one can well imagine, with English loan words and loan translations.

Ich waar in Nei Yarrick gewest. Des iss awwer en wunnerbaari Schtadt. Alleraerscht sehnt mer die Heiser, un viel devun sin heecher ass en Kaerrichetarn. Un do iss es Singergebei. Es iss eenunvazich Schtock hoch, es heechscht Gebei in der Welt.

Alle Schtroosse sin voll Mensche, un all sin in re Hurri. Zuaerscht hawwich gemeent, die Leit deede so renne fer in die Treen, awwer ich hab gfunne, ass es iwwerall so waar, un ich waar schuur, ass sie net all uffem Weg noch der Treen sei kennde.[1]

[1]*Daniel Miller, from "In Nei Yarrick," *Pennsylvania German, *Vol. II.*

In der Schtadt; Es Aerscht Deel

Marriye faahre mer in die Schtadt. Mei gleener
Bruder muss zum Dokder un ich muss zum Zaahdokder.
Des watt awwer ebbes. Nee, mei Bruder iss gaar net
grank; er iss eeggentlich schee gsund. Awwer er geht
im September in die Schul, un en Dokder muss en unnersuche. Un mei Eldre denke, ass mei vedderschde Zeeh
grumm warre. Sie wolle net, ass sie gans schepp waxe.

Gewehnlich iss des fer uns en Blessier, noch der
Schtadt zu geh, awwer wie du der denke kannscht, farricht sich der Billi marriye neizufaahre. Er glaabt,
ass er in der Hassbiddel muss. Er kann sich noch errinnere, ass sei Gremmpaepp mol grank warre iss, un
hot in der Hassbiddel geh misse. Awwer der Gremmpaepp
hot en Hazschlaag ghatt. (Er hot aerscht gedenkt, er
hett yuscht Soodbrenne.)

Un die Memm waar aa mol im Hassbiddel, awwer sie
hot Lungefiewer grickt. (Un sie hot aerscht gemeent,
sie hett en Kalt.) Un yetz(t) denkt der Gleene, ass
ebbes middem arrig letz iss. Er weess, ass wann er
der Huuschde odder Bauchweh odder Halsweh hot, ass er
yuscht im Bett bleiwe muss. Un sogaar wann er Ohreweh
grickt hot, hen mer en graad zum Dokder im Schteddel
genumme. (Mei elderi Schweschder hot mol Hazweh ghatt,
awwer ich denk des iss ebbes schunnscht.)

Un nau muss ich em Billi ausleege, ass nix letz iss,
un ass er net grank iss, un net in der Hassbiddel nei
muss. Mer faahre marriye frieh mit der Maschien nei,
gehne aerscht zum Dokder, hen dann ebbes in em Ress-
d(e)rannt zu esse, laafe wennich in der Schtadt rum
(der Paepp hot Bisness in re Baenk), gehne zum Zaahdokder, un dann widder heem.

Wann mer in die Schtadt faahre, verzeehle mei Eldre
uns wie sie als immer middem Riggelweg faahre hen
misse. Sie hen zum Schteeschen im Schteddel geh misse,
hen sich Faahrzeddel kaafe misse, un sin dann mit der
Dreen noch der Schtadt gfaahre. Die Dreen hot en
Schtieminschein un alles ghatt.

Dann hen sie en Drallielein gebaut, un mer hot aa
mit re Drallie faahre kenne. Ich glaab, ich kann mich
noch draa errinere, awwer ich wees gaar net, wie en
Inschein iewen ausguckt.

In the City; Part One

Tomorrow we are driving into the city. My little
brother must (go) to the doctor and I have to go to the
dentist. That'll be something. No, my brother is not
at all sick; he is actually nice and healthy. But he's
going to school in September, and a doctor has to examine him. And my parents think that my front teeth are
getting crooked. They don't want that they grow all
crooked.

Usually it is a pleasure for us to go into the city,
but as you can think (imagine), Billy is afraid to drive
in tomorrow. He believes he has to go to the hospital.
He can still remember that his grandfather once got sick
and had to go to the hospital. But Granddad had a heart
attack. (He thought at first he just had heartburn.)

And Mom was also in the hospital once, but she got
pneumonia. (And she thought at first she had a cold.)
And now the little one thinks that something is very
wrong with him. He knows that when he has a cough or
stomachache or sore throat that he just has to stay in
bed. And even when he got an earache, we just took
him to the doctor in town. (My older sister had a
heartache once, but I think that is something else.)

And now I have to explain to Billy that nothing is
wrong, and that he is not sick, and doesn't have to go
to the hospital. We'll drive in tomorrow morning with
the car, go first to the doctor, have something to eat
then in a restaurant, walk around in the city a little
(Dad has business in a bank), go to the dentist, and
then home again.

When we drive into the city, my parents will tell us
how they used to have to go with the railroad. They
had to go to the station in town, had to buy (themselves) tickets, and then rode with the train to the
city. The train had a steam engine and everything.

Then they built a trolley line, and one could also
go (ride) with the trolley. I believe I can still remember it, but I don't know at all what an engine even
looks like.

Awwer marriye warre mei Eldre, mei Bruder, un ich
mit der Maschien noch der Schtadt faahre. Mer warre
frieh uffschteh, warre uns schee aaduh, un warre uns
dann frieh uff der Weg mache. Mer warre darrich 's
Schteddel faahre, un dann warre mer uff der Landschtrooss noch der Schtadt faahre. (Ich hoff, es watt
meim Bruder net schlecht warre; des bassiert mannichmol, wann er in re Maschien faahrt.)

Am Schkwaer, wu es Koorthaus un es Raatsgebei
schtehne, watt mei Vadder die Maschien parke. Un dann
warre mer zum Dokder nunner laafe. Er hot sei Affis in
me groosse Affisgebei, uff em zehde Schtock (Floohr).
Un wann mer mit em Elleweeder nufffaahre, dann watt der
Billi sei Maage halde. Ich hab sell aa als geduh.

Dann warre mer waarde misse. Awwer unser Dokder hot
en Buch mit Schtories aus der Biewel, un der Billi
watt's aagucke. Ich weess, er watt die Bilder drin
gleiche. Endlich watt der Dokder em Billi sei Naame
rufe, un mei Memm watt mit em neigeh. Un dann,--well,
du wattscht woll wisse, wie des geht. Die Memm watt en
sich ausziehe helfe misse, un der Dokder watt en messe
un wiege.

Dann watt der Dokder em Billi saage, er soll sei
Maul weit uffschpaare (-schpaerre); un er watt sei Zung
un Zeh un Hals aagucke. Dann watt er em Billi sei Ohre
unnersuche, un sich selwer ebbes in die Ohre schtecke
(ich denk, des Ding heesst en Schteddeskoop odder so
ebbes), un watt em Billi sei Haz un Lunge priefe.
Dann watt der Billi sich umdrehe misse, un der Dokder
watt em en paar Mol mit seim Finger uff der Buckel
schlaage un verleicht aa uff die Bruscht. Dann watt's
verbei sei. Un wann die Rechning kummt, watt der Paepp
mit Kaesch bezaahle--awwer net unne zu brumme.

Vocabulary

    ausleege, ausgelegt: to explain
die Baenk, -s: bank
 es Bauchweh: stomach ache
die Biewel: Bible
 es Bisness: business
die Blessier: pleasure

But tomorrow my parents, my brother, and I will
drive to the city. We will get up early, will get
dressed nicely, and will then start out early. We will
drive through the town, and then we will drive on the
highway to the city. (I hope my brother won't get
[car]sick; that sometimes happens when he rides in a
car.)

At the square, where the courthouse and the city
hall stand, my father will park the car. And then we
will walk down to the doctor. He has his office in a
big office building, on the tenth floor. And when we
go up with the elevator, then Billy will hold his
stomach. I used to do that too.

Then we will have to wait. But our doctor has a
book with stories from the Bible, and Billy will look
at it. I know he will like the pictures in it. Finally, the doctor will call Billy's name, and my mother
will go in with him. And then--well, you probably know
how that goes. Mom will have to help him get undressed
and the doctor will measure him and weigh him.

Then the doctor will tell Billy that he should open
his mouth wide; and he will look at his tongue and
teeth and throat. Then he will examine Billy's ears,
and himself stick something into his ears (I think the
thing is called a stethoscope or something like that),
and will test Billy's heart and lungs. Then Billy will
have to turn over, and the doctor will hit him a few
times on his back with his finger and maybe also on the
chest. Then it will be over. And when the bill comes,
Dad will pay with cash--but not without grumbling.

brumme: to grumble
die Bruscht, Brischt: chest, breast
der Buckel, Bickel (or Buckel): back
der Dokder: doctor
die Drallie: trolley, street car
die Drallielein, -s: trolley line
die Dreen, -s: train
eegentlich: actually

der Elleweeder, -s: elevator
der Faahrzeddel: ticket
    grumm: crooked
    gsund: healthy
der Hals, Hels: throat
 es Halsweh: sore throat
der Hassbiddel: hospital (also Haschbidaal')
 es Haz (Haerz), -er: heart
der Hazschlaag, -schleeg: heart attack
 es Hazweh: heartache
der Huuschde: cough
    iewen: even
die Inschein, -e: engine
 es Kalt: cold
 es Koorthaus, -heiser: court house
die Landschtrooss, -e: highway
die Lung, -e: lung
 es Lungefiewer: pneumonia
der Maage, Meege: stomach
    messe, gemesse: to measure
 es Ohr, -e: ear
 es Ohreweh: earache
    parke: to park
    priefe: to test, examine
 es Raatsgebei, -er: city (town) hall
die Rechning, -e: bill
 es Ressd(e)rannt, -s: restaurant
    schepp: crooked
 es Schkwaer, -s: square
    schlecht warre: to get sick (to the stomach)
der Schteeschen, -s: station
 es Soodbreene(s): heartburn
    uffschparre (-schpaerre): to open wide
    sich umdrehe: to turn (oneself) around
    unnersuche: to examine
    wiege, gewogge: to weigh
    yetz(t): now
der Zaah, Zeeh: tooth
der Zaahdokder: dentist
die Zung, -e: tongue

Vocabulary Notes

es Bauchweh: also es Maageweh, es Leibweh, (die)
    Maageschmatze, Leibschmatze, Bauchschmatze.

Grammar

A. The Future Indicative Tense

1. In the grammar section of Chapter Two, Part Two,
we learned that the PG present tense can be used as an
equivalent of the E future tense, especially if an adverb of future time (marriye, neegscht Woch, and the
like) is used in PG. For this reason the future tense
is rather rare in PG; it is quite possible that there
are more future tense verb forms in this lesson alone
than will be found in scores of pages of PG literature.

Nevertheless, the future tense must be considered a
real part of the PG language, and the fact that its
formation consists of nothing more than the auxiliary
warre (waerre) plus the infinitive of the main verb
makes this an easy tense to learn.

In E the future tense is made up of the auxiliary
will plus the infinitive (without to) of the main verb:
I will go, you will go, we will go, etc. The PG equivalent of "will go" is geh warre (waerre).

Singular	Plural
1. ich wa geh	1. mir,mer warre geh
2. du wascht geh	2. dihr,der watt geh; ihr,er
	warre geh; etc.
3. er,sie,es watt geh	3. sie warre geh

Although the formation of the PG future tense never
varies from the above, word order dictates variations
in the placement of the auxiliary warre:

Normal W.O.	Er watt marriye in die Schtadt faahre.
Inverted W.O.	Marriye watt er in die Schtadt faahre.
Transposed W.O.	Ich weess, ass er marriye in die Schtadt faahre watt.

2. The future tense of a modal auxiliary with a dependent infinitive takes the following forms:

Normal W.O.	Er watt marriye ins Schteddel faahre misse.
Inverted W.O.	Marriye watt er ins Schteddel faahre misse.

Transposed W.O.   Ich weess, ass er marriye ins
                    Schteddel faahre watt misse.

3. As we have noted, the future tense is relatively
rare in PG, and when a PG speaker does use it, he most
likely wishes to express a probability or supposition.
He then frequently adds an expression such as schunn
"already" or woll (wull) "probably," "I suppose":

Mer warre woll (wull) in   We'll probably go into a
  en Ressdrant geh.   restaurant.
Du wattscht woll wisse,   You probably know how
  wie des geht.   that goes.

However, depending on context, "no doubt" could also
be the meaning of woll: "You no doubt know how that
goes."

B. Future Perfect Indicative

1. The "textbook" use of the future perfect tense is
to express an action that is finished prior to another
action in the future: By the time we are back (in the
future), Dad will have paid the bill. The PG equivalent of this E sentence is Bis mer zrick kumme, watt
der Paepp die Rechning bezaahlt hawwe.

Note that the PG future perfect is made up of the
auxiliary warre + the past participle of the main verb
+ the infinitive of the main verb's auxiliary.

Singular	Plural
1. ich wa bezaahlt hawwe	1. mir,mer warre bezaahlt hawwe
2. du wattscht bezaahlt hawwe	2. dihr,der watt bezaahlt hawwe; ihr,er warre bezaahlt hawwe; etc.
3. er,sie,es watt be-zaahlt hawwe	3. sie warre bezaahlt hawwe

2. But the PG speaker uses the future perfect most
often to express probability or supposition, and once
again often adds woll or schunn. Thus ". . . , watt
der Paepp die Rechning schunn bezaahlt hawwe" could express several E equivalents: I suppose Dad will have
already paid the bill; Dad probably will already have

paid the bill; Dad no doubt will already have paid the bill.

3. If a sentence containing the future perfect is used with no other clause containing the future tense, then the future perfect actually refers to past time:

Er watt die Rechning schunn bezaahlt hawwe.	He probably (has) paid the bill already. (In the past, prior to the time the sentence was spoken.)

Again, word order determines the position of the auxiliaries:

Normal W.O.	Der Paepp watt die Rechning be-zaahlt hawwe.
Inverted W.O.	Die Rechning watt der Paepp be-zaahlt hawwe.
Transposed W.O.	Ich weess, ass der Paepp die Rech-ning bezaahlt watt hawwe (watt bezaahlt hawwe).

CHAPTER TWELVE

Part Two

As we continue with our trip to the city, we will find nothing new in grammar and relatively few new vocabulary words. But this lesson ends with a fairly long list of PG vocabulary pertaining to the parts of the body.

Was hot mer in der Schtadt fer Freed?
  'S iss nix ass Leerm un Yacht,
Mer hot kee Ruh de' ganse Daag,
  Kee Schloof die ganse Nacht.

Die Buwe gucke matt un bleech;
  Die Meed sin weiss un dinn;
Sie hen wohl scheene Gleeder aa,
  'S iss awwer nix rechts drin.

Die Schtadtleit sin zu simberlich;
  Sie rege schier nix aa;
Sie brauche net ihr weisse Hend,
  Aus Farricht, 's kummt ebbes draa!

Mir iss zu wennich Grienes do,
  Kee Blumme un kee Beem;
Wann ich en Schtunn im Schteddel bin,
  Dann will ich widder heem.[1]

[1]*Henry Harbaugh, from "Busch un Schteddel,"* <u>*Harfe*</u>.

In der Schtadt; Es Zwett Deel

Graad wie ich immer waar, watt der Billi aa nadierlich froh sei, ass alles in Adder iss, un ass mer dann
esse geh kenne. Mer warre woll (wull) widder in en
Ressdrannt geh, wu es deitschi Koscht gebt. Awwer bis
der Boi uff der Disch kummt, warre mei Bruder un ich
uns schunn satt gesse hawwe.

Mer warre dann iwwer die Schtrooss geh darrefe (awwer mer warre verschpreche misse, uns in Acht zu nemme)
un uns ebbes Siesses kaafe. Mer griege immer en bissel
Geld, fer in der Schtadt zu schpende. Der Gremmpaepp
saagt "Zehrgeld" dezu, awwer ich hab des Watt
schunnscht net gheert. Un bis mer zrick sin, watt der
Paepp die Rechning schunn bezaahlt hawwe--er kann alles
schnell im Kopp ausrechle un braucht ken Aedmaschien--
watt schunn sei Wechsel gezeehlt hawwe, awwer watt noch
am Brumme sei.

Yetz(t) warre mer nunner zum Zaahdokder laafe misse.
Un die Memm watt sich widder die Schappfenschdre aagucke, abbaddich die Gleederschtorfenschdre. Un sie
watt widder saage, ass sie die neie Sitty Faeschens
gaar net gleicht. Awwer mannichmol denk ich, . . .

Dann warre mer widder middem Elleweeder zum siwwede
Schtock faahre misse, un mer warre widder waarde misse,
awwer desmol watt der Dokder mich neirufe, un desmol
watt ich 's Maul uffschparre misse. Un ich hoff, er
watt yuscht en bissel do drin rumfuule, sich die Backe-
zeeh, veddere Zeeh, un es Zaahfleesch aagucke, un mich
widder heemschicke. Awwer wer weess, wie des sei watt.

Bis ich faddich bin, warre mer all schunn genunk
hawwe vun de Schtadtleit, die Kaafer, un die Bisnessleit, wu ihre Bisness in ihre Affissgebeier dreiwe, die
Preise, un iwwerhaabt die Meng Mensche, wu iwwerall(ich)
uff der Schtrooss rumdappe.

Un mer warre all froh sei, wann mer widder deheem
sin, wu der Dreck vum Grund kummt, un net vun der Luft.

In the City; Part Two

Just as I always was, Billy will naturally be happy that everything is in order, and that we can then go eat. We will probably go into a restaurant again where there is German food. But till the pie is on the table, my brother and I will have eaten ourselves full.

We then will be allowed to go across the street (but we will have to promise to watch ourselves) and buy ourselves something sweet. We always get a little money to spend in the city. Granddad calls it spending money, but I have not heard the word otherwise. And till we are back, Dad will already have paid the bill--he can figure out everything in his head fast and needs no adding machine--will already have counted his change, but will still be grumbling.

Now we will have to walk down to the dentist. And Ma will once again look at the shop windows, especially the clothing store windows. And she will say again that she doesn't like the new city fashions at all. But sometimes I think . . .

Then we will have to go to the seventh floor again with the elevator, and we will have to wait again, but this time the doctor will call me in, and this time I will have to open my mouth wide. And I hope that he will just fool around in there a little, look at my molars, front teeth, and gums, and send me home again. But who knows how that will be.

Till I am finished, we will all have had enough of the city people, the customers, and the business people who ply their business in their office buildings, the prices, and especially the crowd of people who are walking everywhere on the street.

And we will all be happy (glad) when we are at home again where dirt comes from the ground and not from the air.

Vocabulary

die Acht: care, caution, heed;   sich in Acht nemme:
      to watch out, to be careful
die Aedmaschien, -e (-s):   adding machine
      ausrechle, ausgerechelt: to figure out
der Backezaah, -zeeh: molar
die Bisnessleit (pl.):   business people
 es Gleederschtorfenschder, -fenschdre: clothing store
      window
      heemschicke: to send home
      iwwerall(ich):   everywhere
      iwwerhaabt: especially
der Kaafer: customer
der Kopp, Kepp: head
die Koscht: food
die Meng, -e: crowd
      neirufe, neigerufe: to call in
der Preis, -e: price
      rumdabbe: to walk around (many times used with the
      connotation of aimless wandering)
      satt: satisfied;   sich satt esse: to eat oneself
      full, to eat enough
 es Schappfenschder, -fenschdre: shop window
die Sitty, -s: city
die Sitty Faeschen, -s: city fashion
      verschpreche, verschproche: to promise
 es Wechsel: change (money)
      woll (wull):   probably, "I suppose," no doubt (see
      grammar)
 es Zaahfleesch: gum(s)
 es Zehrgeld: spending money

DER KOPP, KEPP: HEAD

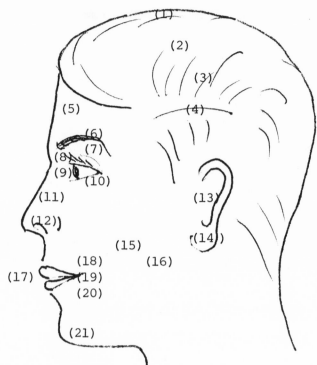

1 der Scheedel, Haern-
  scheedel, die Haern-
  schaal: skull
2 es Haern (Hann): brain
3 die Haar: hair
4 der Scheedel: part
5 die Schtann (Schtaern):
  forehead
6 die Aaggebraue (pl.):
  eyebrows
7 der Aaggedeckel, Schei-
  deckel: eyelid
8 die Aaggehaar: eyelashes
9 der Aaggeabbel: pupil
10 es Aag, Aagge: eye
11 die Naas, Nees: nose
12 es Naaseloch, -lecher:
   nostril
13 es Ohr, -e: ear
14 der Ohrelabbe: ear lobe

15 der Backe, die Wang(e),
   -e: cheek
16 es Gsicht, -er: face
17 es Maul, es Gfress, die
   Gosch: mouth
18 die Lipp, -e; die Lefz,
   -e: lip
19 die Zung, -e: tongue
20 der Zaah, Zeeh: tooth
21 es Kinn: chin
22 der Aademsabbel: Adam's
   apple
23 die Gurgel, der Schluck-
   er, der Schlund, die Kehl:
   throat
24 es Halsreehr: windpipe

242

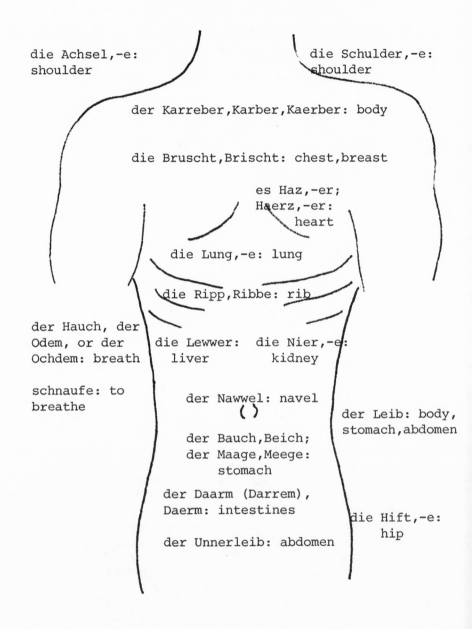

die Achsel,-e: shoulder

die Schulder,-e: shoulder

der Karreber,Karber,Kaerber: body

die Bruscht,Brischt: chest,breast

es Haz,-er; Haerz,-er: heart

die Lung,-e: lung

die Ripp,Ribbe: rib

der Hauch, der Odem, or der Ochdem: breath

die Lewwer: liver

die Nier,-e: kidney

schnaufe: to breathe

der Nawwel: navel

der Leib: body, stomach,abdomen

der Bauch,Beich; der Maage,Meege: stomach

der Daarm (Darrem), Daerm: intestines

die Hift,-e: hip

der Unnerleib: abdomen

DER KARREBER VUN HINNE

die Ankel,
es Gnick, es
Halsgnick: nape
of neck

der Buckel,Bickel: back

es Schulderblaat,-e: shoulder blade

der Rick: back

der Rickschtrang:
backbone,spine,spinal
column

es Greiz: small of
the back

es Blut: blood
der Gnoche: bone
der (die) Muskel,-e:
    muscle
die Haut: skin

ES GLIED, GLIEDER (GLIDDER): LIMB (OF A BODY)

der Aarm (Aarem),-e or Aerm: arm

der Ellbogge: elbow

der Voraarm: forearm

die Fauscht,
Feischt: fist

der Daume:
thumb

die Hand,Hend: hand
der Finger: finger
der Gnechel:
knuckle

es Bee:
leg

der
Schenkel:
thigh

der Fingernaggel,-neggel:
fingernail

es Gnie: knee

die Gniekapp,-e: kneecap
die Gniescheib,-scheiwe:
kneecap

die Waade:
calf

es Schinnbee:
shin(bone)

der Enkel,
der Gnechel:
ankle

der Fuuss,Fiess: foot

der Zehe: toe

der
Faerschde:
heel

der Reie:
instep
die Fuusssohl,-e: sole

We have seen that <u>warre</u> "to become" is the auxiliary of the future tense when used with the infinitive of the main verb. In this lesson we will study the use of <u>warre</u> with the past participle of the main verb to form the passive voice. Read through the text, noting that the past participle goes to the end of a clause or sentence, just as in the case with the perfect tenses. Notice, too, that at times the PG passive voice will have a very poor or even impossible English translation or equivalent. The vocabulary will contain the most common words concerning clothes.

Ich bin glenner wie mei Bruder,
  Un des iss ken Blessier;
Fer alles wu ihm nimmi basst,
  Kummt runner noh zu mir.

Sei Hemd iss v'leicht zu glee am Hals,
  Die Aermel aa zu kaz;
"Ei," saagt die Memm, "es iss en Schand,
  Des schneide fer en Schatz."

"Loss der Bill's noch wennich waere,
  Des geht ihm alsnoch Weil";
Noh kaaft sie dann meim Bruder Sem
  Eens aus der letschte Schteil.

So geht's mit alles wu ich hab,
  Vun Kopp bis unne·naus;
Der Sem grickt immer alles nei,
  Un ich waer's alde aus.[1]

---

[1]*Ralph Funk, from "So Iss Es Ewwe,"* <u>Poems</u>.

Gleeder

Do kann nimman(d) net saage, ass mei Eldre uns Kinner net immer schee gleede. Villeicht draage die yingere die Gleeder, ass die eldere mol ghatt hen, awwer sie basse immer gut un sin immer sauwer. Abbaddich unser Unnergleeder—Unnerhemmer, Unnerhosse, un Unnerreck—sin immer schee frisch gewesche un schee sauwer.

Mer warre immer sogaar mit scheene Schulgleeder in die Schul gschickt (un alle Daag en sauweres Schnuppduch). Awwer mer bhalde sie immer schee, un wisse, ass mer uns rumschtrippe misse, wann mer vun der Schul heemkumme. Un sogaar unser All(e)daagsgleeder un em Paepp un meine Brieder ihre Waerdaagsgleeder sin immer sauwer un schee gflickt.

Un wann mei greessere Brieder aus ihre Iwwerreck, Suuts, un Hosse rauswaxe, dann warre sie vun der Memm fer mich verennert. Der Rock muss gewehnlich net eigenumme warre, weil mei Schuldre beinaah so breet sin wie meine Brieder ihre Schuldre. Awwer mei Aareme (Aerm) un Bee sin net gans so lang, un die Aermel un Hossebee misse nadierlich enwennich kazer gemacht warre. Un wann die Suuts odder Hosse mich nimmi basse—ich wax immer noch, un sie warre ball zu eng un kaz—dann warre sie fer mei gleener Bruder nochemol verennert. Weescht, die Faeschens un Schteils verennere sich net so schnell un net so viel fer die Buwe un Mannsleit.

Un weil mei Fiess beinaah so grooss sin wie meine Brieder ihre Fiess, grick ich ihre Schuh un Schtiwwel, sogaar die Gammschuh un Gammschtiwwel. Mannichmol muss en neier Absatz odder en nei-i Sohl vum Schuhmacher druffgeduh warre, awwer dann sin sie graad so gut wie nei, abbaddich wann sie gebutzt sin warre.

All unsere wollne Schtrimp un Hensching sin vun meine eldere Schweschdere gschtrickt warre. Un wann ich en Loch in me Socke hab, weess ich, ass bis ich en widder aaduh will, watt er woll (wull) widder gschtoppt warre sei. Nadierlich sin unser Hemmer un Blause immer schee gebiggelt; die Koffs un Graage sin abbaddich glatt ausgebiggelt. Un wann alsemol en Gnobb abgerisse watt, watt er vun der Memm glei widder aageneeht.

Un des hot mich yuscht gemaahnt, wie der Billi mol sei Hossesitz gflickt hot. Es waar an me Sunndaag, un

Clothes

No one can ever say that my parents don't always dress us children nicely. Maybe the younger ones wear the clothes that the older ones had once, but they always fit well and are always clean. Especially our underclothes--undershirts, underpants, and slips--are always just freshly washed and nice and clean.

We are even sent to school with nice school clothes (and every day a clean handkerchief). But we always keep them nice, and know that we have to change our clothes when we come home from school. And even our everyday clothes and Dad's and my brothers' work clothes are always clean and nicely mended.

And when my bigger brothers grow out of their overcoats, suits, and pants, then they are altered for me by Mom. The jacket usually doesn't have to be taken in, because my shoulders are almost as broad as my brothers' shoulders. But my arms and legs are not quite so long, and the sleeves and pantslegs must naturally be made a little shorter. And when the suits or pants don't fit me anymore--I am still growing, and they soon get too tight and short--then they are altered again for my little brother. You know, the fashions and styles don't change so quickly and so much for boys and men.

And because my feet are almost as big as my brothers' feet, I get their shoes and boots, even their rubber overshoes and rubber boots. Sometimes a new heel or a new sole has to be put on by the shoemaker, but then they are just as good as new, especially if they have been cleaned.

All of our woolen stockings and gloves have been knitted by my older sisters. And when I have a hole in a sock, I know that by the time I want to put it on again, it will probably have been darned. Naturally our shirts and blouses are always nicely ironed; the cuffs and collars are especially smoothly ironed. And when now and then a button gets torn off, it is immediately sewn back on by Mom.

And that has just reminded me when Billy once mended his pants-seat. It was on a Sunday, and we always go

's watt bei uns Sunndaags immer in die Karrich gange.
Do misse die beschde Sunndaagsgleeder aageduh warre;
yeder muss gut aageduh, gut gedresst, sei. Die Weibsleit misse ihre scheene Fracke waere, un die Mannsleit
misse ihre dunkelblooe Suuts un weisse Hemmer mit
Schlupp aahawwe. (Un yo ken Sackmesser im Hossesack!)

Awwer mir Kinner wisse all, ass die gude Gleeder
ausgezogge warre misse un die All(e)daagsgleeder aageduh warre misse, abbaddich wann mer nausgehne, fer zu
schpiele. Wann's in denne scheene Gleeder gschpielt
watt, warre sie ball versuddelt warre, sell iss gewiss.

An sellem Sunndaag hot awwer der Billi sei gude Hosse aabhalde un iss naus. Un warum er am Geilskeschde-
baam nuffgraddle hot wolle, hen mer noch nie ausgfunne,
awwer er iss so fimf odder sex Fuss nuffkumme--un iss
uff eemol runnergerutscht. Er hot's heere kenne, un
noh hot er's sehne kenne--en groosser Siwweder graad im
Hossesitz.

Der Billi iss nuff in der Schpeicher gschliche, hot
Neez un en Noodel gholt, un hot's Loch zugeneeht. Dann
hot er die Hosse in sei Gleederkammer uffghenkt.

Am neegschde Sunndaag hot die Memm die Hosse rausgenumme, fer sie schee mit re Gleederbaerscht abbaerschde
un dann biggele. . . .

Woche schpeeder hot die Memm uns gsaat, sie hett net
gewisst, eb sie lache odder heile hett wolle. Die Hosse waere gans schee geneeht warre, un sie hett es villeicht gaar net gemarrickt (gemaerrickt). Awwer darrich der weiss Neetz iss es nadierlich gsehne warre,
ass die dunkelblooe Hosse gflickt waare warre!

Vocabulary

 aabhalde, aabhalde: to keep on
 aaduh, aageduh: to put on clothes; sich aaduh:
  to get dressed
 aageduh: dressed
 aaneehe: to sew on
 abbaerschde: to brush off
 abreisse, abgerisse: to tear off
der Absatz, Absetz: heel (as of shoe)
der Aermel, -: sleeve

to church on Sundays (but see grammar). Then the best
Sunday clothes must be put on; everyone must be well
dressed. The women must wear their nice dresses, and
the men must have on their dark blue suits and white
shirts with tie. (And by no means a pocket knife in
our pants pockets!)

But we children all know that the good clothes have
to be taken off and that the everyday clothes have to
be put on, especially when we go out to play. If those
beautiful clothes are played in, then they will soon
get all messed up, that's for sure.

On that Sunday, however, Billy kept on his good
pants and went out. And why he wanted to climb up the
horsechestnut tree, we have never found out, but he got
up about five or six feet--and suddenly slipped down.
He could hear it, and then he could see it--a big seven
right in the seat of his pants.

Billy sneaked up into the second floor, got thread
and a needle, and sewed the hole closed. Then he hung
up the pants in his clothes closet.

On the next Sunday Mom took the pants out, in order
to brush them off nicely with a clothes brush and then
iron them. . . .

Weeks later Mom told us she didn't know whether she
wanted to laugh or cry. The pants had been very nicely
sewn, and she might probably never have noticed. But
through (by means of) the white thread it was naturally
seen that the dark blue pants had been mended!

die All(e)daagsgleeder (pl.): everyday clothes
    ausbiggele: to iron out
    ausfinne, ausgfunne: to find out
    ausziehe, ausgezogge: to take off clothes, to un-
        dress
    sich ausziehe: to undress oneself
    basse: to fit
die Blaus, -e: woman's blouse
    draage, gedraage: to wear; to carry
    druffduh, druffgeduh: to put on
    dunkelbloo: dark blue

einemme, eigenumme: to take in (as a seam in
    clothes)
eng: tight (as clothes); narrow
die Faeschen, -s: fashion
flicke: to mend
der (die) Frack, -e: dress
frisch: fresh
der Gammschtiwwel, -: rubber boot
der Gammschuh, -: rubber overshoe
gebiggelt: ironed
gebutzt: cleaned (polished)
gedresst: dressed
gemaahne: to remind
gflickt: mended
glatt: flat, smooth
gleede: to clothe
die Gleederbaerscht, -de: clothes brush
der Gnobb, Gnebb: button
der Graage, -: collar
 es Hemm, Hemmer: shirt
der Hensching, -: glove
 es Hossebee, -: pant leg
der Hossesack, -seck: pants pocket
der Hossesitz, -e: pants seat
der Iwwerrock, -reck: overcoat
 es Koff, -s: cuff
marricke (maerricke): to notice
nuffgraddle, nuffgegraddelt: to crawl (climb) up
rausnemme, rausgenumme: to take out
rauswaxe: to grow out of
der Rock, Reck: suit coat, jacket
sich rumschtrippe: to change clothes
runnerrutsche: to slip, slide down
schleiche, gschliche: to sneak
der Schlupp, Schlubbe: tie, necktie
 es Schnuppduch, -dicher: handkerchief
die Schteil, -s: style
schtricke: to knit
der Siwweder: the number seven
der Socke, -: sock
die Sohl, -e: sole (of a shoe)
die Suut (es Suut), -s: suit
uffhenke: to hang up (+ direct object)
die Unnergleeder (pl.): underwear
 es Unnerhemm, -er: undershirt

die Unnerhosse (pl.): under pants
der Unnerrock, -reck: petticoat
   verennere: to alter (as clothes)
   versuddle, versuddelt: to soil, ruin, make a mess
     of
die Waerdaagsgleeder (pl.): work clothes
   waere: to wear
   warum?: why?
   woll: wool
   woll (wull): probably
   zuneehe: to sew shut, to sew up

Vocabulary Notes

sich aaduh: also sich aaziehgge, aagzogge.

der Aermel: the sleeve hole is es Aermelloch, -lecher.

sich ausziehgge: also sich ausduh, ausgeduh; sich ab-
   schtrippe.

basse: also fidde.

die Blaus: also der Schmack, die Tschoosi, der Weesd.

einemme: "to let out" is auslosse, ausgelosse.

der (die) Frack: plural also Fracks. Also der (die)
   Gaund, plural Gaund, Gaunde, Gaind, Gainder.

der Gammschtiwwel: also Gum(me)schtiwwel.

der Gammschuh: also der Gum(me)schuh. Also der Iwwer-
   schuh.

der Gnobb: button hole is Gnobbloch, -lecher. zugnibbe
   "to button"; uffgnibbe "to unbutton."

der Graage: also der Halsgraage. Also der Kaller, -s.
   Graage can form compounds with other nouns: Hemmer-
   graage, Rockgraage, etc.

der Hensching: "mittens" are Fauschthensching (literally fist-gloves).

es Koff: also es Breis.

der Rock: A light coat or jacket is also called es
   Kiddel (Kittel).

die (es) Suut: also die (es) Suut Gleeder; der Aazug.

der Unnerrock: A "chemise" is es Weibshemm, -hemmer.

die Waerdaagsgleeder: may or may not be the same as
   die Schaffgleeder "work clothes."

woll: An n is added before endings, e.g., en wollni
   Kapp "a woolen cap."

Grammar

A. The Formation of the Passive Voice

   1. In English, the passive voice is formed with the
various tenses of the auxiliary "to be" and the past
participle of the main verb:

Present	I am seen, you are seen, he is seen, etc.
Past	I was seen, you were seen, etc.
Present perfect	I have been seen, you have been seen, etc.
Past perfect	I had been seen, you had been seen, etc.
Future	I will be seen, you will be seen, etc.
Future perfect	I will have been seen, you will have been seen, etc.

   The passive voice can thus be thought of as being
based on a formula:

       to be + past participle of main verb

The past participle is a constant; no matter what the
person, number, or tense, it remains unchanged. It is
the auxiliary "to be," therefore, that indicates the
tense: am, was, have been, will be, etc.

   2. The preceding can be said of the PG passive voice
also, with the exception that the formula reads:

   warre (to become) + past participle of main verb

The auxiliary warre is the finite verb; it changes from
person to person, number to number, and tense to tense.
The past participle is again a constant; it remains unchanged.

3. The Present Tense of the PG Passive Voice

Singular	Plural
1. ich wa gsehne	1. mir,mer warre gsehne
2. du wascht gsehne	2. dihr,der watt gsehne; ihr,er warre gsehne; etc.
3. er,sie,es watt gsehne	3. sie warre gsehne

4. Passive and Active Counterparts

Let us now see how a passive sentence can be "de-rived" from its active counterpart. In the PG <u>Die</u>
<u>Memm schtoppt der Schtrump</u>, <u>Memm</u> is the subject of the active verb <u>schtoppt</u>. The mother is actively engaged in the action of darning. <u>Schtrump</u> is the direct object of the verb.

In the passive sentence derived from the preceding active sentence, <u>Schtrump</u> becomes the subject (it is now passively being acted upon), and <u>Memm</u> becomes the object of the preposition <u>vun</u> (+ dative). The active verb <u>schtoppt</u> is replaced by the auxiliary <u>warre</u>; and the past participle of <u>schtoppt</u>, the main verb, is placed to the end, in accordance with the usual rules of word order.

Active Voice: Die Memm schtoppt der Schtrump.
Passive Voice: Der Schtrump watt vun der Memm
                gschtoppt.

5. The Various Tenses of the Passive

Now let us combine the two principles explained above. First, passive is a <u>voice</u> and can thus be ren-dered in various tenses. Second, the passive auxiliary <u>warre</u> reflects the tense of the active verb (remember, the past participle is a constant!).

Present: Die Memm schtoppt der Schtrump.
                Der Schtrump watt vun der Memm gschtoppt.

Present   Die Memm hot der Schtrump gschtoppt.
Perfect: Der Schtrump iss vun der Memm gschtoppt
                warre.

Past   Die Memm hot der Schtrump gschtoppt ghatt.
Perfect: Der Schtrump waar vun der Memm gschtoppt
                warre.

Future: Die Memm watt der Schtrump schtoppe.
           Der Schtrump watt vun der Memm geschtoppt
           warre.

Future   Die Memm watt der Schtrump gschtoppt
Perfect: hawwe.
           Der Schtrump watt vun der Memm gschtoppt
           warre sei.

6. Just as the PG future indicative can be used to express probability, just so can the passive form be used, especially if woll (wull) and/or schunn are added:

Future Perfect Active:
Die Memm watt woll der   Mom will probably have
   Schtrump gschtoppt   darned the stocking.
   hawwe.

Future Perfect Passive:
Der Schtrump watt woll   The sock will probably
   schunn vun der Memm   already have been
   gschtoppt warre sei.   darned by Mom.

7. We have seen that the doer of the action, the agent, is the object of the preposition vun: vun der Mudder "by Mom." Two other prepositions--mit (+ dative) "with," and darrich (+ accusative) "through," "by means of"--are also used in PG.

The object of the preposition mit is the instrument with which the action is carried out: Die Hosse warre mit weissem Neez gflickt "The pants were mended with white thread."

The object of the preposition darrich is the means by which the action is carried out: Darrich der weiss Neez iss es gsehne warre, . . . "By means of (through) the white thread it was seen, . . ."

8. Dative Objects

Dative objects of active verbs are retained in passive sentences:

Active: Mer helfe em Paepp.
Passive: Em Paepp watt gholfe.

9. No Active Objects

If the verb in an active sentence has no direct object, then the PG speaker often uses the indefinite subject es in the passive counterpart:

Active: Mer gehne in die Karrich.
Passive: Es watt in die Karrich gange.

But the es is omitted if inverted word order is used or if a question is asked:

Inverted: Sunndaags watt in die Karrich gange.
Question: Watt Sunndaags in die Karrich gange?

Note that the verb reflects the fact that es is a third person, singular form: the verb must be a third person singular form whether or not es is actually used as a subject!

There are, of course, no literal E translations of the above passive sentences with or without es. Depending on context, "one" or "they," or even "I" or "we" could be the subjects of an E active translation (equivalent): One (they) go to church on Sundays, or I (we) go to church on Sundays.

Conversely, an E passive translation often results from a PG active sentence whose subject is the indefinite mer: Mer hot ihre Koches arrig gegliche "Her cooking was liked a lot" (literally, one liked her cooking a lot).

Thus it can happen that two different PG forms, one active and the other passive, can both be translated with E passive: Mer saagt, ass . . . or Es watt gsaat, ass . . . , E "it is said that . . . ."

10. Word Order

The rules of word order of the passive voice are the same as those of the active voice discussed in various previous chapters. But it is worth noting the word order in the following subordinate clauses: the finite verb is not placed last but rather between the past participle and warre.

Present Perfect:
..., abbaddich wann die   ..., especially if the
   Schuh gebutzt sin warre   shoes have been polished

Past Perfect:

..., ass die Hosse   ..., that the pants had
gflickt waare warre.   been mended.

Future:

..., ass bis marriye   ..., that by tomorrow the
der zwedde Hensching   second glove will be
gschtrickt watt warre.   knitted.

11. Passive with Modals

Modals can be used also in the passive voice, in
which case <u>warre</u> becomes the dependent infinitive, and
the rules for double infinitives apply: <u>Der Schtrump</u>
<u>muss gschtoppt warre</u> (present), <u>Der Schtrump hot</u>
<u>gschtoppt warre misse (misse warre)</u> (present perfect),
<u>Der Schtrump watt gschtoppt warre misse (misse warre)</u>
(future).

All rules concerning word order apply; note, for example, the placement of the modal auxiliary (finite
verb) in the following subordinate clauses: <u>Er weess,</u>
<u>ass der Schtrump gschtoppt muss warre; Er weess, ass</u>
<u>der Schtrump gschtoppt hot warre misse</u> (double infini-
tive).

B. The False Passive

German language grammars traditionally mention the
"problem" of the false passive (or statal passive, ap-
parent passive, quasi passive, and other such names).
But the truth of the matter is that these terms refer
to an <u>English</u> language problem! Note that the E verb
"to be" is used in both the translations of the following PG sentences:

Passive: Der Schtrump watt gschtoppt.
           The stocking is darned.

Active: Der Schtrump iss gschtoppt.
           The stocking is darned.

The English language problem derives from the fact
that the verb "to be" is both an auxiliary for the passive voice and a linking (copulative) verb followed by
a predicate adjective. If the predicate adjective happens to be a past participle, then the ambiguity occurs
as above: does the sentence tell us that the sock is

in the process of being darned, or does it tell us that the process of darning is finished and we have a darned sock?

The English speaker many times attempts to solve this problem by using the progressive forms, e.g., The stocking is being darned, was being darned, etc.

Again, no such problem exists in PG. The auxiliary of the passive voice is warre (waerre); no other auxiliary, including sei, can be used in its place to form the passive voice.

CHAPTER FOURTEEN

In PG, very much as in English, the subjunctive mood
is used to express indirect discourse, statements contrary to fact, wishful thinking, and polite questions
or commands (requests). But English subjunctive forms
are very much like indicative forms, and perhaps for
this reason, the average English-speaking person knows
very little about the subjunctive mood, and little
realizes that he is using it at all. The PG speaker is
much more aware of the subjunctive, because the subjunctive forms differ, for the most part, rather markedly
from the indicative. For instance, the E subjunctive
verb form "were" in the clause "If they were here ..."
differs not at all from the past indicative form "were"
in the clause "They were here." But the PG subjunctive
form <u>waere</u> is entirely different from present indicative <u>sin</u> or even the past indicative <u>waare</u>.

The following explanation of the PG subjunctive mood
may at first appear to be rather long-winded, but once
you have worked your way through it, you will know the
satisfaction of having completed the last phase of your
study of Pennsylvania German grammar. And those of you
who choose to go on with PG literature will find this
last phase indispensable.

> *Wann manche wisste,*
> *Was annere gsehne;*
> *Wann annere gengde,*
> *Wu annere gehne;*
> *Wann manche deede,*
> *Was annere duhne;*
> *Dann waer die Welt besser*
> *Fer drin zu wuhne.*[1]

---

[1]*John Birmelin, from "If Many Men Knew,"* PG Verse.

A. Formation of the Subjunctive Mood

1. The PG subjunctive forms are better related to time than to tense, for then we have but two forms to contend with: Present Time Subjunctive to render present and future action or being, and Past Time Subjunctive to render past action or being (any point in time prior to this moment). If we keep in mind the idea of time, then our study of the use of the subjunctive in PG is greatly simplified.

2. Only very few of the total number of PG verbs have their own unique subjunctive forms in the present time. The verbs sei and hawwe are two of these, and because they are so important not only as main verbs but as auxiliaries also, their complete conjugations follow:

a. The Present Time Subjunctive of sei

Singular	Plural
1. ich waer "I would be"	1. mir,mer waere "we would be"
2. du waerscht "you would be"	2. dihr,der waert; ihr, er waere, etc. "you would be"
3. er,sie,es waer "he, she,it would be"	3. sie waere "they would be"

b. The Present Time Subjunctive of hawwe

Singular	Plural
1. ich hett "I would have"	1. mir,mer hedde "we would have"
2. du hettscht "you would have"	2. dihr,der hett; ihr,er hedde; etc. "you would have"
3. er,sie,es hett "he, she,it would have"	3. sie hedde "they would have"

3. The modal auxiliaries all have distinctive Present Time forms, too. The right hand column, labeled Present Time Subjunctive, contains what are actually the 1st and 3rd persons singular forms.

	Present Time
Infinitive	Subjunctive
darrefe, to be allowed to, may	darreft
kenne, to be able to, can	kennt
meege, to care to, wish to	meecht
misse, to have to, must	misst
solle (selle), to be supposed to, should	sott (sett)
wolle (welle), to want to	wott (wett)

4. Seven additional verbs have the following unique Present Time Subjunctive forms:

brauche, to need to	breicht
duh, to do	deet
geh, to go	gingt (gengt)
gewwe, to give	geebt
griege, to get	greecht
kumme, to come	keemt
wisse, to know	wisst

Seven of the preceding verbs--kenne, solle, wolle, duh, geh, gewwe, kumme--are conjugated like kumme, which can be used as a model for all seven:

Singular	Plural
1. ich keemt, "I would come"	1. mir,mer keemde "we would come"
2. du keemtscht "you would come"	2. dihr,der keemt; ihr, er keemde; etc. "you would come"
3. er,sie,es keemt "he, she,it would come"	3. sie keemde "they would come"

The remaining six--brauche, darrefe, griege, meege, misse, wisse--differ from the seven above in that they drop the t of the second person singular ending: du breichscht, darrefscht, greechscht, meechscht, misscht, wisscht.

5. The present time subjunctive of all other PG verbs is formed with the present time subjunctive of the auxiliary duh (deet) and the infinitive of the main verb.

a. Present Time Subjunctive of schwetze (deet schwetze)

Singular	Plural
1. ich deet schwetze "I would talk"	1. mer deede schwetze
2. du deetscht schwetze	2. dihr deet schwetze; ihr deede schwetze; etc.
3. er deet schwetze	3. sie deede schwetze

Remember: all Present Time Subjunctive forms are used to render future time also.

6. The subjunctive mood never changes the rules of word order. Thus like any other auxiliaries (e.g., sei, hawwe, warre), deet is placed in second position or last position, depending on context:

Normal Word Order	Er deet yetzt Deitsch schwetze.
Inverted Word Order	Yetzt deet er Deitsch schwetze.
Transposed Word Order	Wann er Deitsch schwetze deet, . . .

And, as we know, a question would find the verb in first position:

Deet er yetzt Deitsch schwetze?

7. The Past Time Subjunctive

Subjunctive forms in Past Time are used to express action or being at a time prior to the present moment. It is constructed of the Present Time Subjunctive of the auxiliaries hawwe or sei (all rules concerning the choice of auxiliaries also apply in the subjunctive mood), and the past participle of the main verb. This will not seem at all strange if we remember that the past indicative of PG verbs (except sei) is actually made up of the present indicative of an auxiliary plus the past participle of the main verb--in other words, the present perfect tense.

a. An example of the Past Time Subjunctive of a verb whose auxiliary is sei follows:

Singular	Plural
1. ich waer gange "I would have gone"	1. mer waere gange

Singular	Plural
2. du waerscht gange	2. dihr waert gange; ihr waere gange; etc.
3. er waer gange	3. sie waere gange

b. If the auxiliary is <u>hawwe</u>, then the following forms result:

Singular	Plural
1. ich hett ghatt "I would have had"	1. mer hedde ghatt
2. du hettscht ghatt	2. dihr hett ghatt; ihr hedde ghatt; etc.
3. er hett ghatt	3. sie hedde ghatt

8. All rules of word order discussed in Chapter Seven are valid for the Past Time Subjunctive also.

B. Uses of the Subjunctive

1. Indirect Discourse, Present Time

A direct quote (direct discourse) contains all the words originally spoken or written in the same order they were spoken or written. Such a quote normally is enclosed in quotation marks:

Der Tietscher saagt, "Heit iss der aerscht Schuldaag."	The teacher says, "Today is the first school day."

An indirect quote (indirect discourse) is a report of what was said or written:

Der Schulmeeschder hot gsaat, ass heit der aerscht Schuldaag waer.	The teacher said that today was the first school day.

Note, first of all, that no quotation marks are used to enclose an indirect quote; remember that we have only a <u>report</u> of what was said. Secondly, note that PG <u>ass</u> "that" is a subordinating conjunction; it must be preceded by a comma and followed by transposed word order.

But just as E "that" need not be used to introduce

an indirect quote, just so PG <u>ass</u> can be left out:

> Der Schulmeeschder hot
> gsaat, heit waer der
> aerscht Schuldaag.

> The teacher said today
> was the first school
> day.

Note that without <u>ass</u>, the word order is not transposed; we now must follow the rules for normal or inverted word order.

The introduction to a quote has no effect on the
following verb forms of indirect discourse. It can be
present, past, or future; it makes no difference: <u>Er
saagt</u>, <u>hot gsaat</u>, <u>watt saage</u>, <u>ass er gingt</u>. Even the
passive can be used to introduce the indirect quote:
<u>Es watt gsaat, ass er gingt</u>.

2. Simple Declarative Statements

The quote "<u>Heit iss der aerscht Schuldaag</u>" is taken
from page 16 of Chapter One, seventh paragraph. The
direct quotes in this and the following sentence contain verbs that have their own present time subjunctive forms.

> "Dihr wisst, Kinner,
> heit iss der aerscht
> Schuldaag."

> "You know, children, to-
> day is the first
> school day."

> Der Tietscher hot gsaat,
> ass die Kinner wissde,
> heit waer der aerscht
> Schuldaag.

> The teacher said that
> the children knew to-
> day was the first
> school day.

> "Do hawwich die Naame."

> "Here I have the names."

> Der Tietscher hot gsaat,
> ass er die Naame do
> hett (..., er hett die
> Naame do).

> The teacher said that
> he had the names here
> (... said he had the
> names here).

But in the next sentence, <u>kenne</u> "to be acquainted
with" has no unique subjunctive form, and must therefore be used in conjunction with the auxiliary <u>deet</u>:

> "Deel kennich schunn."

> "Some I already know."

> Der Tietscher hot gsaat,
> ass er deel schunn
> kenne deet (..., er
> deet deel schunn ken-
> ne).

> The teacher said that
> he already knew some
> of them.

3. Modal Auxiliaries

The sentence beginning the second paragraph on page 18 contains the modal auxiliary <u>kenne</u> with a dependent infinitive <u>schwetze</u>. No problem; replace the indicative <u>kannscht</u> with the correct form of <u>kennt</u> and obey all the rules of word order.

"Du kannscht Deitsch schwetze."
Der Tietscher hot gsaat, ass die Palli Deitsch
   schwetze kennt (..., die Palli kennt Deitsch
   schwetze).

4. Double Infinitives

The following examples are taken from the explanation of double infinitives on pages 105 and 106 of Chapter Five:

"Ich will im Harrebscht yaage geh."
Er hot gsaat, er wott im Harrebscht yaage geh.

"Ich heer die Schitz schiesse."
Er hot gsaat, er deet die Schitz schiesse heere.

"Die Kinner losse ihre Keits in die Luft geh."
Er hot gsaat, die Kinner deede ihre Keits in
   die Luft geh losse.

5. Interrogatives

a. If an interrogative (question word) is used in direct discourse, then it is used in indirect discourse too. For instance, on page 18 of Chapter One, line 5, the teacher asks, "<u>Wu hockscht du</u>?" If this direct question is asked indirectly, we repeat the interrogative: <u>Der Tietscher hot gfroogt, wu der Tschanni hocke deet</u>. Note the use of transposed word order when an indirect question begins with an interrogative!

b. Then in the next lines the teacher asks, "<u>Gebscht em Meeschder ken Andwatt</u>?" This time, the direct question has no interrogative. We must therefore begin the indirect question with <u>eb</u> (E if, whether), a subordinating conjunction that must be followed by transposed word order: <u>Der Tietscher hot gfroogt, eb er em Meeschder ken Andwatt geebt</u>.

c. And one more example from line 11: "Dutt dei Bruder net Englisch schwetze?" Der Tietscher hot gfroogt, eb ihre Bruder net Englisch schwetze deet. Note that no question mark is used at the end of an indirect question.

6. Commands

Whenever a command in direct discourse is restated in indirect discourse, the imperative form of the verb is replaced by the modal auxiliary sott (sett), but reappears as an infinitive dependent on the modal. All rules of word order concerning modals and dependent infinitives apply. Some examples taken from Chapter Ten:

Ich ruf, "Billi, schteh uff!"

I call, "Billy, get up!"

Ich hab gerufe, ass der Billi uffschteh sott (..., der Billi sott uffschteh).

I called that Billy should get up.

Die Memm ruft, "Sei schtill datt owwe!"

Mom calls, "Be still up there!"

Die Memm hot gerufe, ass ich do howwe schtill sei sott (..., ich sott do howwe schtill sei).

Mom called that I should be still up here.

Die Memm saagt, "Hock dich hie un ess!"

Mom says, "Sit down and eat!"

Die Memm hot gsaat, ass ich mich hiehocke sott un esse (..., ich sott mich hiehocke un esse).

Mom said that I should sit down and eat.

7. Indirect Discourse, Future Time

a. All preceding explanations concerning use of the subjunctive in present time apply to future time as well. Once again, if a verb has its own present time subjunctive form, use it. If not, use the auxiliary deet.

Er saagt, "Ich wa marriye kumme."

He said, "I will come tomorrow."

Er hot gsaat, ass er marriye keemt (..., er keemt marriye).	He said he would come tomorrow.

(By the way, it is not impossible that one hear or see <u>kumme deet</u> or <u>deet kumme</u>.)

Er saagt, "Ich wa marriye ins Schteddel faahre.	He said, "I will drive into town tomorrow."
Er hot gsaat, ass er marriye ins Schteddel faahre deet (..., er deet marriye ins Schteddel faahre).	He said he would drive into town tomorrow.

b. The future subjunctive of a modal auxiliary with a dependent infinitive takes the following form:

Er saagt, "Ich wa marriye ins Schteddel faahre misse."	He said, "I will have to go to town tomorrow."
Er hot gsaat, ass er ins Schteddel faahre misst (..., er misst ins Schteddel faahre).	He said that he would have to drive into town tomorrow.

8. Indirect Discourse, Past Time

If direct discourse makes use of a tense indicating action or being in the past (prior to this moment), then the indirect discourse is rendered with the Past Time Subjunctive. The following examples are taken from Chapter Eight.

Er saagt, "Mei Freind iss widder verbeikumme."	He says, "My friend came by again."
Er hot gsaat, ass sei Freind widder verbeikumme waer (..., sei Freind waer widder verbeikumme).	He said that his friend had come by again.
Er saagt, "Er hot noch nie so en Kochoffe gsehne ghatt."	He says, "He had never before seen such a stove."
Er hot gsaat, ass er noch nie so en Kochoffe	He said that he had never before seen such a stove.

```
gsehne hett (..., er
hett noch nie so en
Kochoffe gsehne).
```

9. Contrary-to-Fact (Conditional) Sentences

In E, a sentence expressing an idea that is contrary
to fact usually contains two parts, a subordinate
clause beginning with if and the main clause (sometimes,
but not often, beginning with then). The if-clause
sets up a condition: If we had the money (on the condition that we had the money), . . . The following
main clause (then-clause) states a conclusion to the
preceding condition: If we had the money, we would buy
a house. Of course, it is also possible to reverse the
clauses: We would buy a house if we had the money.

The preceding can be said of PG also. The PG equivalent of "if" is wann; the PG equivalent of "then" is
dann (also do), but it is not used very often. (E
"then" PG dann must never be used if the main clause
precedes the subordinate clause.)

As with direct discourse, it is best that we think
in terms of time, present and past. We then need only
match Present Time Subjunctive to present (future) time
and Past Time Subjunctive to past time.

a. Present Time

Present time refers to right now. If we had the
money (right now), we would buy a house (right now).
Present time dictates the use of Present Time Subjunctive in PG: Wammer (wann + mer) 's Geld hedde, deede
mer en Haus kaafe.

As with indirect discourse, the PG speaker uses subjunctive forms unique to certain verbs:

Wann ich Geld hett, gengt ich ins Schted- del.	If I had money, I would go into town.
Wanner (wann + er) die Zeit hett, keemt er mit uns.	If he had the time, he would come with us.

Note that the main clause employs inverted word
order because it follows the subordinate clause.

The clauses may be reversed: <u>Mer deede en Haus</u>
<u>kaafe, wammer 's Geld hedde</u>; <u>Er keemt mit uns, wanner</u>
<u>die Zeit hett</u>. Note that the main clause reverts to
normal word order when it is placed first in the sentence. Of course, a subordinate clause has transposed
word order no matter where it appears in a sentence.

The subordinating conjunction <u>wann</u> may be dropped,
but the finite verb must then stand in first position,
followed immediately by the subject: <u>Hedde mer 's</u>
<u>Geld, deede mer en Haus kaafe</u>. (<u>Mer deede en Haus</u>
<u>kaafe, hedde mer 's Geld</u> must be considered rare.)

b. Future Time

Whatever has been said of present time is true of
future time also. Very often an adverbial expression
of future time is used: <u>Wann er die Zeit hett, keemt</u>
<u>er marriye mit uns</u>.

c. Past Time

Past time refers to time prior to this moment in
time. In E, such time is rendered with the perfect
forms: If I had had the money (prior to this moment),
I would have bought a house (prior to this moment).
In PG, we use the Past Time Subjunctive: <u>Wann ich es</u>
<u>Geld ghatt hett, hett ich 's Haus kaaft</u>; <u>wann er die</u>
<u>Zeit ghatt hett, waer er mit uns kumme</u>.

Very often an adverbial expression of past time is
used: <u>Wanner geschder die Zeit ghatt hett, waer er</u>
<u>mit uns kumme</u>.

Again, the clauses may be reversed: <u>Er waer mit uns</u>
<u>kumme, wanner die Zeit ghatt hett</u>.

And <u>wann</u> may be dropped: <u>Hett er (hedder) die Zeit</u>
<u>ghatt, waer er geschder mit uns kumme</u>.

10. "Wishful Thinking"

In E, we often use the subjunctive--together with
"only"--to express a wish that we know or believe will
not come true. Such a wishful thought can be pictured
as nothing more than the <u>if</u>-clause of a contrary-to-
fact sentence. The PG equivalent of "only" is <u>yuscht</u>.

Present Time (Future):

Wann er yuscht do waer!	If only he were here!
Wammer yuscht 's Geld hedde!	If only we had the money!

Past Time:

Wanner yuscht do gwest waer!	If only he had been here!
Wammer yuscht 's Geld ghatt hedde!	If only we had had the money!

The conjunction <u>wann</u> may be omitted, resulting in the need for inverted word order: <u>Waer er yuscht do!</u> <u>Waer er yuscht do gwest!</u> <u>Hedde mer yuscht's Geld!</u> <u>Hedde mer yuscht 's Geld ghatt!</u>

The subjunctive may be used after a wish expressed literally with the verb <u>winsche</u> "to wish":

Ich winsch, ass sie do waere (..., sie waere do).	I wish they were here.
Ich winsch, sie waere do gwest.	I wish they had been here.

11. The Subjunctive after <u>ass (wie) wann</u> "as if," "as though."

The subjunctive is used after <u>ass wie wann</u> because what follows is evidently contrary to fact (or why else would we say "as if" or "as though"?).

Er schwetzt, ass (wie) wann er viel Geld hett.	He talks as if he had a lot of money.
Er hot geschwetzt, ass wann er mol viel Geld ghatt hett.	He talked as if he once had had a lot of money.

Note that <u>ass (wie) wann</u> is followed by transposed word order.

12. Polite Questions and Requests

In E, we many times use the subjunctive mood to ask for something in a polite way. Instead of "Do you have change?" we ask, "Would you have change?" The PG equivalents would be, "Hoscht Wechsel?" and "Hettscht Wechsel?"

Indicative:
Kannscht mer saage, waer   Can you . . .
    des iss?

Subjunctive:
Kennscht mer saage, waer   Could you . . .
    des iss?

Likewise, we often "soften" a command by using the subjunctive. We can tell someone, "Do me a favor!" But we can also use the subjunctive to request more politely, "Would you do me a favor." PG equivalents would be "Duh mer en Gfalle!" and "Deetscht mer en Gfalle duh."

As a matter of fact, these polite commands very often take on the sound of a question, as if we were giving the person spoken to the option of saying "No": "Deetscht mer en Gfalle duh?" "Would you do me a favor?"

VOCABULARY

ALLE SARDE WARDE

Warde, Warde, so viel Sarde!
Deel vun do un deel vun darde;
Deel, die sin so wiescht verdreht
As sie niemand meh verschteht;

Graade, grumme, g'scheite, dumme,
Nemmt sie ewwe wie sie kumme;
O, wie hot's doch so viel Sarde
Pennsylvanisch Deitsche Warde![34]

A

aa: also, too (1,1)
aabaue, aagebaut: to build on (add to) (8,1)
aabhalde, aabhalde: to keep on (13,1)
aabinne, aagebunne: to tie up to (11,1)
der Aabinnposchde: hitching post (11,1)
aaduh, aageduh: to put on, get dressed (4,1)
sich aaduh: to dress (oneself) (10,1)
aafange, aagfange: to begin (10,1)
aageduh: dressed (13,1)
sich (nanner) aagreische, aagegrische: to yell at (each other) (10,1)
aagucke, aageguckt: to look at (10,1)
aahawwe, aaghatt (aaghadde): to have on, wear (clothes) (9,1)
aaneehe, aageneeht: to sew on (13,1)
der Aar(e)m, -e: arm (9,2)
aasehne, aagsehne: to look at (6,1)
abbaddich: especially (1,2)
abbaerschde, abgebaerscht: to brush off (13,1)
der Abbeditt: appetite (4,2)
abbedittlich: appetizing (10,2)
der Abbel, Ebbel: apple (9,1)
abdrickle, abgedrickelt: to dry off (something) (6,1)
sich abdrickle, abgedrickelt: to dry oneself off (10,1)
der Abdritt: privy (8,1)
abfaahre, abgfaahre: to drive off (11,1)
abhacke, abghackt: to chop off (6,1)
abhelle, abgehellt: to clear (as after a storm) (5,2)
abkehre, abkehrt: to sweep off (10,1)
abmeehe, abgemeeht: to mow down (9,1)
abreisse, abgerisse: to tear off (13,1)
abrobbe, abgerobbt: to pluck (9,1)
der Absatz, Absetz: heel (of shoe) (13,1)
abschpiele, abgschpielt: to rinse (wash) off (6,1)
abschrecke, abgschrocke: to scare off (9,1)
abschtaawe, abgschtaabt: to dust off (6,2)
abseege, abgseegt: to saw off (6,1)
abwische, abgewischt: to wipe off (5,2)
achtgewwe, achtgewwe: to watch out, be careful (6,2)
die Adder: order (5,1)
in adder: in order (5,1)
die Aedmaschien, -e (-s): adding machine (12,2)
aensere: to answer (1,2)
die Aent, Aents: aunt (2,1)
die Ant, Ants: aunt (2,1)

die Aerbs, -e: pea (9,1)
sich aergere: to get angry (10,1)
der Aermel: sleeve (13,1)
es Aermelloch, -lecher: sleeve hole (13,1)
die Aern: harvest (9,2)
aerscht: first (1,1)
die Aerwet, Aerwedde: work (8,2)
sich an die Aerwet mache: to get to work, start in working (10,1)
aeryeds (aryeds): somewhere (5,2)
es Affiss: office (12,1)
es Affissgebei, -er: office building (11,1)
der Aldweiwersummer: Indian summer (5,2)
all(e): all (3,2)
alle beed: both (3,2)
alle Daag: every day (1,2)
alle zwee: both (all two) (3,2)
die All(e)daagsgleeder (pl.): everyday clothes (13,1)
allee: alone (11,1)
der Allerheilechedaag: All Saints Day (5,1)
alles: everything (1,1)
allgebott: now and then (1,2)
als: be in the habit of, "used to" (4,2)
als noch: still, yet (4,2)
alsemol: sometimes, at times, now and then (1,2)
alt: old (2,2)
der Alt: old man (4,2)
am (an + em): at, to the (dative) (1,1)
der Ameerigaaner: American (5,1)
die Amschel, Amschle: robin (5,2)
an: at, to (1,1)
andwadde: to answer (1,2)
die Andwatt, Andwadde: answer (3,1)
annere: other(s)
arrig: very, (terribly) (1,2)
aryeds (aeryeds): somewhere (5,2)
aus: out of (6,1)
ausbiggele, ausgebiggelt: to iron out (13,1)
ausfinne, ausgfunne: to find out (13,1)
ausgraawe, ausgegraawe: to dig out (9,1)
auslege, ausgelegt: to explain (12,1)
ausmischde, ausgemischt: to clean a stable (10,1)
ausrechle, ausgerechelt: to figure out (12,1)
sich ausruhe, ausgeruht: to rest (10,2)
ausrutsche, ausgerutscht: to slip (5,2)
ausschpanne, ausgschpannt: to unhitch a horse (10,1)
ausschteche, ausgschtoche: to dig out (as potatoes) (9,1)
ausse: outside (6,1)
vun ausse: from the outside (6,1)
auswennich: "by heart" (3,1)
auswennich lanne: to memorize (3,1)
sich ausziehe, ausgezoge: to undress (10,2)
awwer: but (1,1)
die Ax, Ex (Axe): ax (8,1)

es Baad: bath (6,2)
 baade, gebaadt: to bathe (10,2)
 sich baade: to bathe oneself
 (10,2)
die Baad(e)schtubb, -e (-schtuwwe):
 bathroom (6,2)
der Baadezuwwer, -ziwwer: bathtub
 (6,2)
 baalkebbich: baldheaded (2,2)
der Baam, Beem: tree (5,2)
der Baamgaarde: orchard (9,1)
 baarfiessich: barefooted (5,2)
der Baare (Barre): loft of barn (8,2)
der Baart, Beert: beard (2,2)
es Babbelmaul, -meiler: gossip,
 blabbermouth (2,2)
 babble, gebabbelt: to talk; chat,
 gossip (4,2)
es Babier, Babiere: paper (1,1)
 backe: to bake (2,2)
es Backes: baking (2,2)
der Backeschtee: brick (6,1)
es Backeschteehaus, -heiser: brick
 house (6,1)
der Backezaah, -zeeh: molar (12,2)
der Backoffe, -effe: bake oven (6,1)
 badde, gebatt: to help, avail
 (11,1)
 baddere: to bother (5,1)
die Baenk, -s: bank (12,1)
die Baerge: birch (9,2)
es Baergehols: birchwood (9,2)
 ball: soon (2,1)
der Ball, Balle: ball (5,2)
 sich balwiere, gebalwiert: to
 shave oneself (10,1)
der Balwierer: barber (11,2)
die Bank, Benk: bench (1,1)
der Barr (Baer), -e: bear (11,1)
der Barrick, Barrigge: mountain (4,2)
die Barschtubb, -schtuwwe (-schtubbe):
 barroom (4,1)
der Baschdert: wet field used for
 pasture (9,1)
 basse: to fit (13,1)
 bassiere, (ge)bassiert: to happen (5,1)
es Bauchweh: stomach ache, belly
 ache (12,1)
 baue: to build (8,1)
es Bauerehaus, -heiser: farmhouse
 (6,2)
die Bauerei, -e: farm (1,2)
die Bauereleit (pl.): farmers, farm
 people (2,2)
die Baueremaschien, -e: farm machine
 (8,2)
die Bauersfamilye: farm family (2,2)
es Bechli, -n: brook (8,1)
 sich bedrachde, bedracht: to
 look at (observe) oneself (10,1)
es Bee: leg (10,1)
der Beeddaag: Thanksgiving (5,1)
 beede, gebeedt: to pray (10,2)
der Beerebaam (Bierebaam), -beem:
 pear tree (9,1)
es Beewi, Beewis: baby (2,1)
 begraawe, begraawe: to bury (11,
 2)
 bei: at, with (2,1)
es Beik, -s: bike (11,1)
es Beil, -e: hatchet (8,1)
 beinaah: almost, nearly (5,2)

der Bein(d)baam, -beem: pine tree
 (5,1)
es Beiseckel, -s: bicycle (11,1)
 beisse, gebisse: to bite (9,1)
die Bell, Belle (Bells): bell (1,1)
der Belsnickel: Santa Claus (5,1)
der Bensil, Bensils: pencil (1,1)
der Bese(m): broom (10,1)
der Bese(m)schtiehl: broom handle
 (10,1)
 bescht: best (2,2)
es Bett, Bedder: bed (4,1)
 ins Bett geh: to go to bed (4,1)
die Bettdeck, -decke: bedcover (6,2)
es Bettduch, -dicher: bed sheet
 (6,2)
die Bettschtubb, -e (-schtuwwe):
 bedroom (6,2)
die Bettzeit: bedtime (4,2)
 bezaahle: to pay (4,1)
der Bezaahlsdaag: pay day (4,1)
 bhalde, bhalde: to keep (1,2)
der Bicherschank, -schenk: book case
 (6,1)
 sich bicke: to bend down (10,2)
es Biefschteek, -s: beef steak (8,1)
die Biewel: Bible (12,1)
der Bierro, Bierros: bureau (6,2)
es Biggelbord: ironing board (10,2)
 biggele: to iron (10,2)
es Biggeleise: flat iron (10,2)
es Bild, Bilder: picture (1,2)
 bis: to, till; as far as (3,2)
es Bisness, -e: business (12,2)
die Bisnessleit (pl.): business
 people (12,2)
es Blaeckbord, -borde: blackboard
 (1,1)
 blaffe: to bark (5,2)
 blaffich: "barky," said of dog
 in habit of barking (10,1)
die Blans, -e: plant (9,1)
es Blatt (Blaat), Bledder: leaf;
 page (5,1)
die Blaus, -e: woman's blouse (13,1)
der Bleibensil, Bleibensils: lead
 pencil (1,1)
 bleiwe, gebliwwe: to remain,
 stay (1,2)
die Blessier: pleasure (12,1)
es Blettli, Blettlin: saucer (6,1)
 bliehe: to bloom, blossom (5,2)
der Blitz: lightning (5,2)
 blitze: to lighten, lightening
 (5,2)
 blodd: bald (2,2)
 bluuge, gebluugt: to plow (9,2)
die Blumm, -e: flower (5,2)
der Bo, Bos: beau, boyfriend (2,1)
es Bobbel, Bobbelcher: baby (2,2)
der Bodde(m): floor, ground (1,1)
es (der, die) Boggi, -s (or -):
 buggy (8,2)
der Boi, -s: pie (2,2)
die Bortsch, -e: porch (6,1)
der Boss: kiss (10,2)
 bosse: to kiss
 brauche: need, have need of (4,1)
es Breckfescht: breakfast (4,2)
der Breckfeschtdisch: breakfast table
 (4,2)
es Breddich, -e: sermon (11,2)
 breddiche: to preach (11,2)
 breet: broad, wide (6,1)
der Brei: porridge, pap (10,1)

es Brennhols: firewood (6,1)
der Brief, Briefe: letter (4,2)
der Briefdreeger: letter carrier,
    mailman (4,2)
die Brieh: juice (9,1)
der Briehdroog, -dreeg: scalding
    trough (8,1)
    bringe, gebrocht: to bring (8,2)
es (der, die) Briwwi: privy (8,1)
    brode, gebrode: to fry (8,1)
die Brotwaerscht, -e: sausage (8,1)
der Bruder, Brieder: brother (1,1)
es Bruderskind, -kinner: nephew,
    niece (2,1)
    brumme: to grumble (12,1)
der Brunne: well (8,1)
die Bruscht, Brischt: chest, breast
    (12,1)
der Bsuch: company, visitor(s) (4,1)
    uff Bsuch geh: to go visiting
    (4,1)
    Bsuch griegge: to get company
    (4,1)
    bsuche: to visit (4,1)
der Bu, Buwe: boy (1,2)
es Buch, Bicher: book (1,1)
der Buchschtaab, -schtaawe: letter
    (of alphabet) (1,2)
    buchschtaawiere: to spell (1,2)
der Buckel, Bickel: back (12,1)
der Budder: butter (8,2)
es Budderfass, -fesser: butter churn
    (8,2)
die Bump, Bumbe: pump (8,1)
es Bumphaus, -heiser: pump house
    (8,1)
der Bungert: orchard (9,1)
es Buro, Buros: bureau (6,2)
der Busch, Bisch: woods, forest (5,1)
der Butzemann, -menner: scarecrow
    (9,1)
der Butzlumbe: dust rag, cleaning rag
    (10,2)

                    D

der Daadi, -s: daddy, father (2,1)
der Daag, -e: day (1,1)
die Daag(es)helling: dawn (4,2)
    daagelang: for days (at a time)
    (9,2)
es Daageslicht: daylight (6,1)
es Daal, Deeler: valley (5,2)
der Daaler: dollar (3,1)
es Dach, Decher: roof (5,2)
es Dachfenschder, -e: dormer window
    (skylight) (6,1)
der Dankdaag: Thanksgiving (5,1)
    danke: thanks (1,2)
es Dankfescht: Thanksgiving (5,1)
    dann: then (1,1)
    dann un wann: now and then (1,2)
    danse: to dance (4,1)
die Dansschtubb, -schtubbe (-schtuwwe):
    dance hall (11,1)
    darr (daerr): dry (5,2)
    darrich: through (1,2)
    datt (dadde): there (1,1)
    datthieschtelle, datthiegschtellt:
    to place (stand) over there
    (9,1)
der Debbich, - or -e: quilt, (bed-
    spread) (6,2)
die Deck, Decke: ceiling (1,1)
der Deckel: lid, cover (6,1)

    deel: some (1,1)
    defor: for it (6,1)
    degeege: against it (6,1)
    deheem: at home (1,2)
der Deich, Deiche: pond (5,2)
    deide, gedeit: to point (3,2)
    deier: expensive (2,2)
    Deitsch: German (1,1)
der Deller: plate (6,1)
    demarriye: this morning (1,2)
der Dengelschtock, -schteck: anvil on
    which scythe is sharpened by
    hammering (9,1)
    dengle, gedengelt: to sharpen a
    German scythe by hammering (9,1)
    denke: to think (2,2)
    denn: (see vocabulary note, 1,2)
    dennoh: after that (5,2)
    dernoh: after that (4,2)
    des: that (3,2)
der Desk (Dest), -s: desk (1,1)
    desmol: this time (1,2)
    deswegge (deswege): therefore
    (4,1)
    dick: thick, fat (5,2)
    dief: deep (5,2)
die Dier, Diere: door (1,1)
der Diergnobb, -gnebb: doorknob (6,1)
es Dierli, -n: gate (9,1)
die Dierschwell, -e: door sill (6,1)
    dimm(e)le, gedimmelt: to thunder
    (in the distance) (5,2)
die Dinde, ink (1,1)
es Ding, Dinger: thing (4,2)
    dinn: thin (6,2)
der Disch, Dische: table (1,1)
der Dischlappe: dishcloth (6,1)
der Dischlumbe: dishrag (6,1)
    do: here (1,1)
die Dochder, Dechder (Dochdere):
    daughter (2,1)
der Dochdersmann, -menner: son-in-
    law (2,1)
    dohin: in here (6,1)
der Dokder: doctor (12,1)
    dositze, dogsotze (dogsesse): to
    sit here (10,2)
    dot: dead (2,1)
    dotmache, dotgemacht: to kill
    (10,2)
    draageh, draagange: to begin do-
    ing something, to start in (9,1)
    draage, gedraage: to wear; to
    carry (13,1)
der Draeckder, -s: tractor (8,2)
der Draemp, -s: tramp (10,1)
die Drallie, -s: trolley, street car
    (12,1)
die Drallielein, -s: trolley line
    (12,1)
die Draub, Drauwe: grape (9,2)
    draue: to trust (2,2)
    draus: outside (3,2)
    draushalde, drausghalde: to keep
    someone or something out (9,1)
die Drauwerank, -e: grape vine (9,2)
    drehe: to churn butter (8,2); to
    turn (3,2)
    dreckich: dirty (10,2)
die Dreen, -s: train (12,1)
    drei: three (1,1)
die Drepp, Drebbe: step, stair (6,1)
    dresche, gedrosche: to thresh
    (9,2)
die Dreschdenn: threshing floor of
    barn (8,2)

die Dreschmaschien, -e: threshing
machine (9,2)
drin: in there (6,1)
drinke, gedrunke: to drink (4,2)
drinsitze, dringsotze (dringsesse):
to sit in (it) (6,1)
dritt: third (3,1)
zu dritt: the three of us, you,
them (3,1)
driwwe: over (there) (4,2)
der Drobbe: drop (5,2)
der Droht, Dreht: wire (9,1)
der Droog, Dreeg: water trough (10,1)
drowwe: above, up (there) (4,2)
drucke: dry (5,2)
druff: on (upon) it (1,2)
druffduh, druffgeduh: to put on
(13,1)
druffleige, druffgelegge: to lie
on (it) (6,1)
druffmoole, druffgemoolt: to
draw, paint on (11,1)
druffschteh, druffgschtanne: to
stand on (it) (6,1)
es Drunnelbett, -bedder: trundle bed
(6,2)
duh, geduh: to do (1,2)
die Dullebuhn, -e: tulip (6,2)
dumm: stupid (1,1)
sich dumm(e)le, gedummelt: to
hurry (10,1)
der Dummkopp, -kepp: stupid person
(10,1)
dunkel: dark (3,2)
dunkelbloo: dark blue (13,1)
es Dunkes: gravy (10,2)
der Dunner: thunder (5,2)
dunnere: to thunder (5,2)
duschber (duschder): dusk (4,2)

E

eb: before (3,1)
der Ebbelbaam, -beem: apple tree
(5,2)
der Ebbelboi, -s: apple pie (9,1)
ebber: someone (1,2)
ebbes: something (1,2)
ebbes schunnscht: something else
(11,2)
eblang: before long (4,2)
die Eck, Ecke: corner (1,2)
Eckball: corner ball (a game)
(1,2)
der Eckschank, -schenk: corner cupboard (6,1)
ee: one (1,1)
der Eeche: oak (9,2)
es Eecheholts: oak wood (9,2)
es Eechel: acorn (9,2)
die Eeg, Eege: harrow (9,2)
eegen: own (8,2)
eegentlich: actually (12,1)
es Eel: oil (6,2)
eent: one (of them) (3,1)
efder(s): often (8,2)
ehrlich: honest, honorable (2,2)
eifense, eigfenst: to fence in
(9,1)
eifrich: diligently, busily (8,2)
die Eil, -e: owl (10,2)
eimache, eigmacht: to preserve
(e.g., fruit) (9,2)
einemme, eigenumme: to take in
(as a seam in clothes) (13,1)

es Eis: ice (5,2)
eischlaage, eigschlaage: to
strike (lightning) (5,2)
eischpanne, eigschpannt: to
hitch up a horse (10,1)
eischteige, eigschtigge: to
climb in (get on) a vehicle
(11,1)
eise (eisn + ending): iron (6,1)
es Eise: iron (8,1)
eisner: of iron (8,1)
elder: older (1,2)
die Eldre (pl.): parents (2,1)
eldscht: oldest (2,1)
elf(e): eleven (2,2)
der Elleweeder, -s: elevator (12,1)
es End: end (6,2)
eng: tight (as clothes); narrow
(6,2)
Englisch: anyone not PG (11,1)
der Enkel, -n: grandson (2,1)
ennichebber: anybody (4,2)
ennihau: anyhow (2,2)
enwennich: a little (2,2)
die Ent, Ende: duck (9,2)
erkenne, erkennt: to recognize
(1,2)
sich errinnere (aerrinnere): to
remember (11,1)
es: it (3,2)
die Esch: ashes (6,2)
der Eschemittwoch: Ash Wednesday
(5,1)
esse, gesse: to eat (5,1)
die Esssache: foodstuff (6,1)
der Essschank, -schenk: pantry (6,1)
die Essschtubb, -e (-schtuwwe): dining room (6,1)
etliche (edliche): some, a few
(1,2)
ewwerscht: uppermost, topmost
(6,1)

F

faahre, gfaahre: to drive, ride
in a vehicle (4,1)
es Faahrraad, -redder: bicycle
(11,1)
der Faahrzeddel: ticket (12,1)
faddich: finished, ready (4,2)
die Faeckdri, -s: factory (2,2)
die Faeschen, -s: fashion (13,1)
falle, gfalle: to fall (5,2)
falsch: false, wrong (5,1)
es Falschgesicht, -gsichter: false
face, mask (5,1)
die Familye: family (2,1)
die Fanness, (Farness): furnace (6,2)
die Farreb, Farrewe: color (5,1)
farrichde (faerrichde), gefar-
richt: to fear (5,2)
sich farrichde (faerrichde): to
be afraid, to fear (10,2)
die Faasenacht: Shrove Tuesday (5,1)
fascht: almost (5,2)
die Faasezeit: Lent (5,1)
der Fassant, Fassande: pheasant (5,1)
fattyaage, fattgeyaagt: to chase
away (11,1)
faul: lazy (10,2)
die Fedder, Feddere: feather, pen
(13,1)
die Fedderdeck, -e: feather bed (6,2)
feege, gfeegt: to sweep (10,1)

feege, gfeegt: to sweep (10,1)
der Fehler: mistake (1,2)
der Feierdaag, -daage: holiday (5,1)
feiere (feire), gefeiert: to celebrate (5,1)
es Feierhaus, -heiser: firehouse
(11,2)
der Feierheerd: fireplace (6,1)
der Feierhund, -e: andiron (6,1)
die Feierinschein, -e: fire engine
(11,2)
die Feierkump(e)ni, -s: fire company
(11,2)
der Feiermann, Feierleit: fireman
(8,2)
der Feiervoggel, -veggel: lightning
bug (10,2)
die Feierwaerricks: fireworks (5,1)
es Feld, -er: field (5,2)
die Fens, -e: fence (6,1)
es Fenschder, Fenschdere: window
(1,1)
die Fenschderscheib, -scheiwe: window
pane (6,1)
die Fensemaus, -meis: chipmunk (9,2)
der Fenseriggel: fence rail (8,1)
fer: for (1,2)
ferwas?: why? (4,2)
es Fescht, Feschde: feast, festival
(5,1)
der Fettkuche: doughnut (5,1)
es Fettkichelche, -r: doughnut (5,1)
fied(e)re, gfiedert: to feed
(animals) (10,2)
figgere: to figure (do arithmetic) (1,2)
es Fillsel: filling (10,2)
fimf(e): five (2,2)
finne, gfunne: to find (9,2)
sich nanner finne: to find each
other (10,2)
finschder: dark (4,2)
der Fisch, -e: fish (9,2)
fische: to fish (11,1)
die Fleddermaus, -meis: butterfly
(9,1)
flicke: to mend (10,2)
fliege, gflogge: to fly (5,1)
fliesse, gflosse: to flow (9,1)
die Flint, Flinde: gun (5,1)
der Flohr: floor, story of a building
(6,1)
die Foohn, -s: telephone (4,2)
die Fraa, Weiwer: wife, woman (rare)
(2,1)
der (die) Frack, -e: dress (13,1)
die Frantschtubb, -e (-schtuwwe):
front room (6,1)
der Freind, -e: friend (4,1)
die Freindschaft: relations, relatives (2,2)
fresse, gfresse: to eat (of animals) (9,1)
frieh: early (1,2)
der Friehyaahrsdaag, -e: spring day
(5,2)
friere, gfrore: to freeze (5,2)
frisch: fresh (13,1)
frischgelegt: freshly laid (8,2)
froh: happy, glad (5,2)
die Froog, -e: question (1,2)
frooge, gfroogt: to ask (1,1)
der Frosch, Fresch: frog (9,2)
der Froscht: frost (5,2)
der Fuddergang, -geng: passageway
along stalls to facilitate feed-
ing barn animals (8,2)

funkle: to sparkle (4,1)
fuule: to fool (1,2)
der Fuuss, Fiess: foot; unit of
measurement (8,2)

G

gaar net: not at all (2,2)
gaar ken: none at all (2,2)
der Gaarde, Geerde (also Gaarde):
garden (6,1)
die Gaardefens, -e: garden fence
(8,1)
gaardegraawe: to dig in the garden (9,1)
die Gaardehack, -e: garden hoe (9,1)
es Gaardesach, -e: garden produce
(9,1)
es Gacki: child's word for egg
(10,1)
gaerdle, gegaerdelt: to garden
(8,1)
gaern: like (2,2)
der Gammschtiwwel, -e: rubber boot
(13,1)
der Gammschuh: rubber (over)shoe
(13,1)
der Gang, Geng: hall, hallway (6,1)
die Gans, Gens: goose (5,2)
der Gaul, Geil: horse (8,2)
gauze: to bark (5,2)
die Gawwel, Gaww(e)le: fork (6,1)
gaww(e)le, gawwelt: to fork,
to pitch (as with a hay fork)
(9,1)
gaxe: to cackle (8,2)
es Gebei, Gebeier: building (6,2)
gebiggelt: ironed (13,1)
geborni: nee (2,2)
der Gebottsdaag, -daage: birthday
(5,1)
gebrode: fried (8,2)
gebutzt: cleaned, polished (13,1)
gedarrt (gedaerrt): dried (10,2)
der Gedechtnissdaag: Memorial Day
(5,1)
gedresst: dressed (13,1)
die Geelrieb, -riewe: carrot (9,1)
gegge (geege): toward (against)
(5,2)
gegge Nadde: toward the north
(5,2)
gegge Sudde: toward the south
(5,2)
es Geglebber: clattering (10,2)
geh, gange: to go (1,1)
danse geh: to go dancing (4,1)
schaffe geh: to go to work
(4,1)
es Geilche: little horse, pony (10,
2)
es Geilsgscharr (-gschaer): harness
(10,1)
der Geilsposchde: hitching post (11,
1)
der Geilsschtall, -schtell: horse
stall, stable (8,2)
der Geizhals: stingy person (2,2)
es G(e)lennder: bannister (6,2)
gell?: right? (6,1)
gemaahne: to remind (13,1)
gemaescht: mashed (10,2)
die Gemee: congregation (11,2)
es Gemeensglied, -er: member of a
congregation (11,2)

es Gemies: vegetables (9,1)
genunk: enough (3,1)
es Gerabbel: rattling (10,2)
gerooscht: roasted (10,2)
geschder: yesterday (4,1)
die Geschwischder (pl.): siblings,
brothers and sisters (2,2)
es Gewaerz: spices (6,1)
gewehnlich: usually (3,2)
es Gewidder: thunderstorm (5,2)
die Gewidderrud: lightning rod (5,2)
der Gewidderschtarrem, -e: thunderstorm (5,2)
die Gewidderwolk, -e: thundercloud
(5,2)
gewiss: of course, certainly
(1,1)
gewwe, gewwe: to give (2,2)
es gebt: there is, there are
(4,1)
gflickt: mended (13,1)
die Giesskann, -e: sprinkling can
(10,1)
es Giwwelend, -enner: gable end
(6,1)
es Glaas, Glesser: glass (6,1)
der Glaasschank, -schenk: china
closet, cabinet for glassware
(6,1)
glaawe, (ge)glaabt: to believe
(2,2)
glatt: flat, smooth (13,1)
glee (gleen + ending): small
(1,1)
gleede, gegleedt: to clothe
(13,1)
die Gleeder: clothes (4,1)
die Gleederbaerscht, -de: clothes
brush (13,1)
die Gleederkammer: closet (6,2)
der Gleederschank, -schenk: clothes
cabinet, wardrobe (6,2)
es Gleederschtorfenschder, -fensch-
dre: clothing store window
(12,2)
die Gleene (pl.): the little ones
(e.g., the children) (8,2)
glei: immediately, right now
4,2)
gleiche, gegliche: to like (1,2)
glingele: to ring (4,2)
glitzerich: glittery (4,2)
es Glofder: cord (of wood) ((,2)
der Gnobb, Gnebb: button (13,1)
der Gott: God (4,1)
Gott sei Dank: thank God (4,1)
es Graab, Greewer: grave (11,2)
der Graabschtee: grave stone (11,2)
graad: exactly, right; straight
(3,2)
der Graage: collar (13,1)
der Graahne: spigot (6,1)
die Graasmaschien, -e: lawnmower
(8,1)
graawe, gegraawe: to dig (8,1)
die Grabb, -e: crow (9,1)
graddle, gegraddelt: crawl (9,2)
grank: sick (1,2)
sich gratze: to scratch oneself
(10,2)
der Grautkopp, -kepp: cabbage head
(9,1)
es Grautsellaat: cole slaw (10,2)
greesermache, greesergemacht: to
make larger (6,1)
es Greewie: gravy (10,2)

die Greid: chalk (crayon) (1,1)
der Greizweg, -e: crossroads (11,2)
der Greizwegschtor, -s: store at
crossroads (11,2)
die Gremmemm, -s: grandmother (2,1)
der Gremmpaepp, -s: grandfather (2,2)
die Grick, -e: creek (9,1)
griege, grickt: to get, receive
(4,1)
griesse: to greet (1,2)
der Grisch(t)baam, -beem: Christmas
tree (5,1)
der Grisch(t)daag: Christmas (5,1)
es Grisch(t)kindel: Christmas gift,
present (5,1)
der Grisch(t)munet: December, the
Christmas month (5,1)
grooss: large, big (1,1)
der Groossdaadi, -s: grandfather
(2,1)
es Groossdaadihaus: addition to
farmhouse for grandparents (8,1)
die Groosseldre: grandparents (2,1)
die Groossmammi, -mammis: grandmother (2,1)
die Groossmemm, -s: grandmother (2,2)
grumm: crooked (12,1)
die Grummbeer (-bier), -e: potato
(9,1)
der Grummbeereausmacher: potato digger (9,1)
der Grummbeereschtengel, -schtengle:
potato stalk (9,1)
grummbucklich: hump-backed (2,2)
der Grundflohr: ground floor (of
building) (6,1)
der Grundkeller: ground cellar (8,1)
die Grundsau, -sei: groundhog (5,1)
es Gscharr (Gschaerr): dishes (6,1)
gscheit: smart (1,1)
es Gschenk, -e: present, gift (5,1)
gschmookt: smoked (8,2)
es Gschwall, -s: squirrel (9,2)
gschwischich: between (6,1)
es Gsicht, Gsichder: face (5,1)
gsund: healthy (12,1)
gucke: to look, peer (1,2)
es Guckbox: television (11,2)
Gude Nacht: good night (10,2)
der Gumme: gum (12,2)
die Gummer, -(e): cucumber (9,1)
der Gummeresellaat, -e: cucumber
salad (9,1)
gut: good, well (1,1)
gutguckich: good looking (2,2)

H

es Haar, -e: hair (2,2)
der Haas, -e: rabbit (5,1)
hacke, ghackt: to chop, to hack;
to hoe (5,1)
der Hackglotz, -gletz: chopping
block (8,1)
der Haffe, Heffe: pot (6,1)
der Haggel: hail (5,2)
haggele, ghaggelt: to hail (5,2)
halb: half (4,2)
die Halbnacht: midnight (3,2)
halbwegs: halfway (3,2)
halde, ghalde: hold; stop (9,2)
der Hals, Hels: throat (12,1)
es Halsweh: sore throat (12,1)
halwer: half (3,2)
die Hand, Hend: hand (1,2)

es Handduch, -dicher: towel (6,1)
der Hassbiddel: hospital (12,1)
hatt: hard, difficult (2,2)
die Hausaerwet: housework (10,2)
die Hausdier, -e: house door (6,1)
es Hausrot: furniture (6,2)
die Haussache (pl.): house furnishings (6,2)
der Hausschlissel: house key (6,1)
hawwe, ghatt or ghadde: to have
(2,1)
der Hawwer: oats (9,2)
es Haz, -er: heart (12,1)
der Hazschlaag, -schleeg: heart attack (12,1)
es Hazweh: heartache (12,1)
die Heckefens, -e: fence made of
brush (9,1)
heem: home (4,2)
die Heemet, Heemede: home (6,1)
heemgemacht: home made (8,1)
heemgereest: home raised (9,1)
heemkumme, heemkumme: to come
home (4,2)
heemschicke, heemgschickt: to
send home (12,2)
heere, gheert: to hear, listen
(1,2)
heess: hot (5,2)
heesse: to call, name; be called,
named (1,1)
heidesdaags: these days (9,1)
hei(e)re, gheiert: to marry (2,1)
heile, gheilt: to cry (11,2)
es Heisel: little house (8,1)
heit: today (1,1)
helfe, gholfe: to help (2,2)
die Helft: half (3,2)
hell: light, bright (3,2)
die Hembeer, -e: raspberry (9,2)
es Hemm, -er: shirt (13,1)
hendich: handy (5,2)
henge, ghanke: to hang (1,1)
der Hensching: glove (13,1)
hewe, ghowe: to lift (1,2)
es Hickerihols: hickory wood (8,1)
sich hiehocke, hieghockt: to sit
(oneself) down (10,1)
sich hieleege, hiegeleegt: to lie
down (10,2)
sich hiesetze, hiegsetzt: to sit
down (10,2)
der Himmel: sky, heaven (4,1)
es Hinkel: chicken (8,2)
es Hinkelfudder: chicken feed (8,2)
es Hinkelhaus, -heiser: chicken
house (8,2)
die Hinkelschtang, -e: chicken roost
(8,2)
der Hinkelwoi: chicken hawk (9,2)
hinne: in back, in the rear (1,2)
hinnedraus: out in back (6,2)
hinnenausgeh, hinnenausgange: to
go out back (6,1)
der Hinnerhof, -heef: back yard (6,1)
hinnich: behind, in back of (6,1)
die Hitz: heat (5,2)
hiwwe: on this side (8,2)
die Hochzich: wedding (2,1)
die Hochzichdripp: wedding trip (2,1)
die Hochzichrees: wedding trip (2,1)
hocke, ghockt: to sit (1,1)
hoffe, ghofft: to hope (5,2)
es Hoi: hay (8,2)
die Hoigawwel, -le: hay fork (9,1)
der Hoischreck, -e: grasshopper (9,2)

der Hoischtock, -schteck: hay stack
(8,2)
der Hoiwagge, -wegge: hay wagon (9,1)
es Hols: wood (6,1)
es Holsheisel: woodshed (8,1)
der Holskarreb: basket for carrying
wood (8,1)
die Holskischt, -e: woodbox (6,1)
es Holsland, -lenner: woodland (9,2)
der Holsoffe, -effe: wood stove (6,1)
hooch: high (5,1)
es Hoochschtielche, -r: little high
chair (6,1)
der Hoochschtuhl, -schtiehl: high
chair (6,1)
die Hosse (pl.): pants (10,2)
es Hossebee: pant leg (13,1)
der Hossesack, -seck: pants pocket
(13,1)
es Hossesackwedder: very cold
weather (5,2)
der Hossesitz: pants seat (13,1)
howwe: up (6,2)
do howwe: up here (6,2)
der Hund, -e: dog (4,2)
hunne: down (here) (6,2)
es Hunswedder: bad weather (5,2)
hupse, ghupst: to hop (5,2)
es Hutschel (Hutsch), -cher: colt
(8,2)
der Huuschde: cough (12,1)

I

die Iem, -e: bee (10,2)
die Iemens (Umens), -e: ant (10,2)
iesi: easy (1,2)
iewen: even (12,1)
immer: always (1,2)
immer noch: still, yet (2,2)
in: in, into (1,1)
ins (in + es): into the (+ accusative) (1,1)
die Inschein, -e: engine (12,1)
iwwer: over (1,2)
iwwerall(ich): everywhere (12,2)
iwwerhaabt: especially (9,2)
iwwerich: left over, remaining
(3,1)
iwwermarriye: day after tomorrow
(4,1)
der Iwwerrock, -reck: overcoat (13,1)

K

der Kaader: tom cat (10,2)
kaafe, kaaft: to buy (4,2)
der Kaafer: customer (12,2)
die Kaart, Kaarde: card (5,1)
es Kaendi: candy (2,2)
der Kaerpet (Karrepet), -s: carpet
(6,1)
es Kaesch: cash (11,2)
der Kaffi: coffee (4,2)
die Kaffikann, -e: coffee pot (6,1)
es Kalb, Kelwer: calf (5,1)
der Kall (Kaerl), -s: fellow, chap,
guy (2,2)
kalt: cold (5,1)
es Kalt: cold (12,1)
der Kandel: spouting (6,1)
es Kappibuch, -bicher: copybook
(1,2)

die Karrich (Kaerrich), -e: church
    (4,1)
der Karrichetarn, -e: church steeple,
    tower (11,1)
der Karrichhof, -heef: cemetery (2,2)
der Kaschebaam (Kaerschebaam), -beem:
    cherry tree (9,1)
die Katz, -e: cat (6,1)
der Katzefisch: catfish (11,1)
    kaue, kaut: to chew (10,2)
die Kaundi, -s: county (2,2)
    kaz: short (5,1)
der Kees: cheese (8,2)
der Keffer: bug (10,2)
die Keit, -s: kite (5,1)
    kehre, kehrt: to sweep (10,1)
der Keller: cellar (6,2)
die Kellerdier, -e: cellar door (6,2)
    es Kellerloch, -lecher: cellar door-
    way (leading to outside) (6,2)
der Kellerschlaag: cellar door (out-
    side) (6,2)
die Kellerschteeg, -schteege: cellar
    stairs (6,2)
die Kelt: cold (5,2)
    ken, kee: no, not a, none (1,2)
    kenne, kennt: to be able to, can
    (1,1)
    kenne, kennt: to be acquainted
    with (1,1)
der Kessel: kettle (6,1)
die Kett, -e: chain (10,1)
der Kettehund, -e: dog kept chained
    up (10,1)
    es Ketzel, -cher: kitten (10,1)
    es Ketzli, -n: kitten (10,1)
die Kich, -e: kitchen (6,1)
die Kichedier, -e: kitchen door (6,1)
der Kichedisch, -e: kitchen table
    (6,1)
    es Kichelche, Kichelcher: little
    cake (5,1)
der Kicheschank, -schenk: kitchen
    cupboard (6,1)
    kiehl: cool (5,2)
der Kiehschtall, -schtell: cow stall,
    stable (8,2)
    es Kind, Kinner: child (1,1)
die Kinnerwieg, -e: cradle (6,2)
    es Kinskind, -kinner: grandchild
    (2,1)
die Kischt, Kischde: box, chest (6,1)
der Kissel: sleet (5,2)
    es Kisselwedder: sleety weather (5,2)
    kissle, kisselt: to sleet (5,2)
der Kiwwel: pail (10,1)
die Klass, -e: class (1,1)
    koche, kocht: to cook (2,2)
    es Koches: cooking (2,2)
der Kochoffe, -effe: stove (6,1)
    es Koff, -s: cuff (13,1)
der Kohloffe, -effe: coal stove (6,1)
der Kolwe: ear of corn, cob (9,2)
    es Koorthaus, -heiser: court house
    (12,1)
der Kopp, Kepp: head (1,2)
    es Koppche, Koppcher: cup (4,2)
    es Kopp(e)kisse: pillow (6,2)
    es Koppli, Kopplin (Kopplicher): cup
    (6,1)
    es Koppweh: headache (5,1)
    koschde, koscht: to cost (11,2)
die Koscht: food (12,2)
der Kuche: cake (2,2)
die Kuh, Kieh: cow (8,2)
    kumme, kumme: to come (1,1)
der Kuss, Kiss: kiss (10,2)

L

    es Laab: foliage (5,1)
der Laade, Leede: shutter (6,1)
    laafe, geloffe: to walk (1,2)
    es Land, Lenner: land (1,1)
    uff em Land: in the country
    (1,1)
die Landschtrooss, -e: highway (12,1)
    lang: long (1,2)
    lanne: to learn (1,1)
der Lattwarrick (Lattwaerrick): apple
    butter (9,1)
    laut: loud(ly) (1,2)
die Leckschen, -s: election (5,1)
der Leckschendaag: election day
    (5,1)
der Lecktrickoffe, -effe: electric
    stove (6,1)
    leddich: single, unmarried (2,2)
die Leeder, -e: ladder (8,2)
der Leederwagge, -wegge: (hay)wagon
    with rack (9,1)
    leer: empty (9,2)
der Leffel: spoon (6,1)
    leigge, gelegge: to lie (be re-
    cumbent) (2,2)
    es Leinduch, -dicher: bedsheet (6,2)
die Leit (pl.): people (2,2)
    lengermache, lengergemacht: to
    make longer (6,1)
    lese, gelese: to read (1,1)
    letz: wrong (1,2)
    lewe, gelebt: to live (2,1)
    es Lewesmiddel: food (6,1)
die Lewesschtubb, -e (-schtuwwe):
    living room (6,1)
die Lewwerwa(er)scht, - or -e: liver-
    wurst (8,1)
    es Licht, -er: light (5,2)
    lieb: dear (4,2)
    es Lied, -er: song (11,2)
    liewe, geliebt: to like, love
    (2,2)
    es Liftel: breeze (5,2)
    links: (to the) left (1,1)
    es Loch, Lecher: hole (5,1)
    loofe: to loaf (1,2)
    losmache, losgemacht: to loosen,
    to untie (11,2)
    losse: to let, allow (5,1)
die Luft: air (5,1)
die Lung, -e: lung (12,1)
    es Lungefiewer: pneumonia (12,1)

M

der Maage, Meege: stomach (12,1)
    mache: to make, do (1,2)
die Maddratz, -e: mattress (6,2)
    es Maebelhols: maple wood (9,2)
    es Maentel(bord): mantel (6,1)
die Maetsch: match (2,2)
die Maetschgeil (pl.): matched
    horses (8,2)
die Mammi, -s: mommy (2,1)
der Mann, Menner: man, husband (2,1)
    mann(i)chmol: sometimes (5,2)
der Mannskall, -leit: man, fellow
    (2,2)
    marricke (maerricke): to notice
    (13,1)
der Marriye: morning (4,1)
    marriye: tomorrow (1,1)
    marriyeds: mornings, in the
    mornings (1,2)

es Marriyedsesse: breakfast (4,2)
es Marriye-esse: breakfast (10,1)
der Marriyeschtann, -e: morning star (4,2)
die Maschien, -e (-s): car (8,2); machine (9,1)
die Mauer, -e: wall (8,2)
es Maul, Meiler: mouth (9,2)
  maule: to mouth, grumble (11,1)
der Medizienschank, -schenk: medicine cabinet (6,2)
es Meedel, Meed: girl (2,2)
  meege, gemeecht: to care to, may (5,1)
  meehe: to mow (9,1)
die Meehmaschien, -e (or -s): mower (9,1)
der Meelmann, -menner: mailman (4,2)
  meene: to mean; to think, be of the opinion (2,2)
die Meening, -e: opinion (6,1)
der Meeschder: school master (1,1)
  meh, mehner: more (2,1)
  meinde: to mind, to remember (5,2)
  melke, gemolke: to milk (10,1)
der Melkeemer: milk pail (10,1)
der Melkschtuhl, -schtiehl: milking stool (10,2)
die Memm, -s: mom (2,1)
die Meng: crowd (12,2)
  die Meng Mensche: crowd of people (12,2)
  menscht: most (1,2)
  mer: one, they, people (3,1)
  messe, gemesse: to measure (12,1)
es Messer, Mess(e)re: knife (6,1)
die Mickedier, -e: screen door (6,1)
der Middaag, -e: midday (3,2)
es Middaagesse: noonday meal, lunch (4,2)
die Middaagsschtunn: noon hour (4,2)
die Middernacht: midnight (3,2)
die Middnacht: midnight (3,2)
  mied: tired (9,2)
die Miehl, -e: mill (11,1)
der Miehldeich, -er: mill pond (11,1)
es Miehlraad, -redder: mill wheel (11,1)
es Miehlrees: mill race (11,1)
der Miller: miller (11,1)
die Millich: milk (4,2)
die Millichboddel, -boddle: milk bottle (4,2)
der Millichmann, -menner: milkman (4,2)
  minanner, (mitnanner): together (1,2)
der Minuddezoiyer: minute hand (3,2)
die Minutt, Minudde: minute (3,2)
der Mischt: manure (10,1)
die Mischtgawwel, -gawwle: manure fork (10,1)
der Mischthaahne: barnyard (manure pile) rooster (10,1)
der Mischthaufe: manure pile (10,1)
  misse, gemusst: must, to have to (3,1)
  mit: with (1,1)
  mithelfe, mitgholfe: to help along (10,1)
  mitnemme, mitgenumme: to take along (8,2)
  mitschpiele, mitgschpielt: to play with (10,2)
  mittwegs: midway (4,1)
die Moiblumm, -e: May flower (5,2)
  mol(1): once, at some time (1,2)

der Mo(o)nd: moon (4,2)
  sich mucke: to move (10,2)
die Mudder, Midder: mother (2,1)
der Muddersdaag: Mother's Day (5,1)
die Mudderschprooch, -e: mother tongue (1,2)
der Munet, -ede: month (5,1)
die Muschkratt, -radde: muskrat (9,2)
es Muviehaus, -heiser: movie house (11,2)
die Muvies: movies (11,2)

N

der Naame: name (1,1)
die Nacht, Nachde (Nechde): night (3,2)
es Nachtesse: supper (4,2)
der Nachthaffe, -heffe: chamber pot (4,2)
es Nachthemm, -hemmer: night shirt (4,2)
  nadierlich: naturally (1,2)
der Nammidaag (Nummidaag), -e: afternoon (4,1)
  nammidaags: in the afternoon, afternoons (4,1)
der Nattwind (Nardwind): north wind (5,2)
der Nascht, Nescht: branch (6,1)
  nau: now (3,1)
  naus: out (1,2)
  nausdraage, nausgedraagt: to carry out (6,2)
  nausfiehre, nausgfiehrt: to lead out (6,1)
  nausfliege, nausgflogge: to fly out (10,1)
  nausgeh, nausgange: to go out (6,1)
  nausyaage, nausgeyaagt: to chase out (8,2)
  neddemol(1), netemol(1): not even (2,2)
  nee: no (1,1)
  neegscht: next, near (4,2)
die Neehern: seamstress (10,2)
es Neehes: sewing (10,2)
der Neez: thread (10,2)
  nei: new, in (1,2)
  neibeisse, neigebisse: to bite into (something) (9,1)
  neibumbe, neigebumbt: to pump in (8,1)
  neiduh, neigeduh: to put in (6,1)
  neifaahre, neigfaahre: to drive in (11,1)
  neifiehre, neigfiehrt: to lead into (6,1)
  neigeh, neigange: to go in(to) (6,1)
  neirufe, neigerufe: to call in (12,2)
  neischlaage, neigschlaage: to strike (lightning) (8,2); to knock or hit into (10,1)
  neischtecke, neigschteckt: to stick in(to) (6,1)
  neischtelle, neigschtellt: to place in(to) (8,1)
  nemme, genumme: to take (1,2)
es Nescht, Neschder: nest (8,2)
  net: not (1,1)
  newwich: beside, next to (1,2)
  nie: never (1,1)

niemand (nimmand): no one (1,2)
nimmi: no longer (2,1)
nix: nothing (1,1)
noch: yet, still (2,1)
noch net: not yet (3,1)
der Nochber, -e: neighbor (4,2)
die No(o)chberschaft, -e: neighborhood (1,2)
noh: then (3,1)
nooch: after, according to (3,2)
noochrufe, noochgerufe: to call
after (11,2)
noochschmeisse, noochgschmisse:
to throw after (11,2)
die Noodel, Noodle: needle (10,2)
nucke: to nod (1,2)
nuffgawwle, nuffgawwelt: to fork
up, to pitch up (9,1)
nuffgeh, nuffgange: to go up (6,1)
nuffgraddle, nuffg(e)raddelt: to
crawl (climb) up (13,1)
nuffhupse, nuffghupst: to jump up
(6,1)
nuffschteige, nuffgschtigge: to
climb up (8,2)
die Nummer, -e: number (3,1)
nunner: down (6,2)
nunnerbrenne, nunnergebrennt: to
burn down (8,2)
nunnergeh, nunnergange: to go
down (6,2)
die Nuss, Niss: nut (9,2)

O

es Obscht: fruit (5,1)
odder: or (1,1)
der Offe, Effe: oven, stove (1,2)
es Offerohr, -e: stove pipe (6,1)
oft: often (2,2)
oftmols: often (1,2)
es Ohr, -e: ear (12,1)
es Ohreweh: ear ache (12,1)
es Oi, Oier: egg (8,2)
der Onkel (Unkel), -s (also -): uncle
(2,1)
es Oschderbonnet, -s: Easter bonnet
(5,1)
die Oschdere (pl.): Easter (5,1)
der Oschderhut, -hiet: Easter hat
(5,1)
es Oschderoi,-oier: Easter egg (5,1)
der Oschdersunndaag: Easter Sunday
(5,1)
oweds: evenings (3,2)
die Owetluft: evening air (4,2)
der Owetschtann, -e: evening star
(4,2)
owwe: above (1,1)
es (die) Owwerdenn, -er: hay loft
(8,2)

P

paar: couple, few (6,1)
der Paepp: father, "pap," "pop" (2,1)
die Pann, -e: pan (6,1)
der Pannhaas: scrapple (8,1)
parke: to park (12,1)
der Parlor, -s: parlor (6,1)
der Parre: preacher, pastor (11,2)
der Paschingbaam (Paerschingbaam),
-beem: peach tree (9,1)
der Peffer: pepper (6,1)

es Pikter, Pikters: picture (1,2)
pischbere: to whisper (1,2)
es Pitscher, -s: pitcher (6,1)
der Poschde: post (8,1)
der Poschtmeeschder: postmaster
(11,2)
der Preis, -e: price (12,2)
priefe: to test, examine (12,1)

R

es Raad, Redder: wheel, bicycle
(11,1)
der Raahm: cream (8,2)
es Raatsgebei, -er: city (town) hall
(12,1)
rabble, gerabbelt: rattle (4,2)
die Rassel, Rassle: rattle (9,2)
die Rasselschlang, -e: rattlesnake
(9,2)
die Ratt, Radde: rat (10,2)
raus: out (4,2)
rauskumme, rauskumme: to come out
(6,1)
rausnemme, rausgenumme: to take
out (12,1)
rausschpritze, rausgschpritzt: to
squirt out (9,1)
sich raussuche, rausgsucht: to
pick out for oneself (11,2)
rauswaxe: to grow out of (13,1)
der Reche: rake (8,1)
rechle, gerechelt: to reckon, do
mathematics (1,2)
die Rechning, -e: bill (12,1)
recht: right (1,2)
rechts: (to the) right (1,1)
reese: to raise (9,2)
der Regge: rain (5,2)
der Reggeboge: rainbow (5,2)
der Reggedrobbe: rain drop (5,2)
reggere: to rain (5,2)
reibringe, reigebrocht: to bring
in (6,1)
reide, geridde: to ride (10,2)
reif: ripe, mature (9,1)
der Reife: frost (5,2)
reihole, reigholt: to bring in
(6,1)
reikumme, reikumme: to come in
(6,1)
der Reitgaul, -geil: riding horse
(8,2)
renne: run (1,2)
es Ressd(e)rannt, -s: restaurant
(12,1)
die Retsch: gossip (4,2)
richtich: right (5,2)
der Riggelweg,-e: railroad (2,2)
ringe, gerunge: to ring (1,1)
es Rinsfleesch (Rindfleesch): beef
(10,2)
rischde: to prepare (as food)
(10,1)
robbe: to pick, pluck (9,1)
der Rock, Reck: suit coat, jacket
(13,1)
rode: to guess (10,2)
der Rogge: rye (9,2)
die Rotrieb, -riewe: red beet (9,1)
rufe, gerufe: to call (4,2)
ruffbumbe, ruffgebumbt: to pump
up (8,1)
ruffrufe, ruffgerufe: to call up
(10,1)

der Ruh(k)daag: day of rest (4,1)
  ruhich: quietly (10,2)
  rum: around (1,2)
  rumbatsche, rumgebatscht: to
    splash around (10,2)
  rumdappe, rumgedappt: to walk
    around (12,2)
  rumdrehe, rumgedreht: to turn
    around (6,1)
  sich rumdrehe, rumgedreht: to
    turn around (10,1)
  rumgeh, rumgange: to go around
    (11,1)
  rumgraddle, rumgegraddelt: to
    climb around (8,2)
  rumgucke, rumgeguckt: to look
    around (6,2)
  rumhocke, rumghockt: to sit
    around (4,1)
der Rumleefer: tramp (10,1)
  rumschpringe, rumschprunge: to
    run around (5,2)
  sich rumschtrippe, rumgschtrippt:
    to change clothes (13,1)
  runnerfalle, runnergfalle: to
    fall down (10,2)
  runnerrutsche, runnergerutscht: to
    slip down (13,1)
  runslich: wrinkly, wrinkled (2,2)

                    S

  saage: to say (1,1)
die Sache (pl.): things (8,1)
der Sack, Seck: bag, sack (5,1)
der Saddel, Seddel: saddle (10,2)
der Sals: salt (6,1)
der (die) Sank (Sangk): sink (6,1)
  satt: satisfied (12,2)
  sich satt esse: to eat enough
    (12,2)
die Satt, Sadde: sort, kind (5,1)
die Sau, Sei: pig (8,2)
der Sauergraut: sauerkraut (9,1)
  sauwer: clean (6,1)
der Schaawer: scraper (8,1)
der Schadde: shadow (5,2)
der Schaddebaam, -beem: shade tree
    (6,1)
der Schaffdaag, -e: work day, weekday
    (4,1)
  schaffe: to work (2,2)
der Schaffgaul, -geil: workhorse (8,2)
die Schaffgleeder (pl.): work clothes
    (4,1)
die Schaffleit (pl.): workers, work-
    ing people (4,1)
der Schank, Schenk: cabinet, cupboard
    (6,1)
der Schannschtee: chimney (6,1)
der Schapp, -s: shop (11,2)
 es Schappfenschder, -fenschdre: shop
    window (12,2)
  scharre (schaerre): to scratch
    (8,2)
die Schaufel, -le: shovel (8,1)
  schee (adv.): nice(ly) (3,1)
  schee (scheen + ending): nice,
    beautiful (1,2)
die Scheer, -e: scissors (10,2)
der Scheerm: umbrella (5,2)
die Scheier, Scheire: barn (5,2)
die Scheierbrick: approach to thresh-
    ing floor of a bank barn (8,2)
 es Scheierdach, -decher: barn roof
    (5,2)

 es (die) Scheierdenn, -er: barn
    floor (9,2)
 es Scheierdor: barn door (8,2)
der Scheierhof, -heef: barnyard (8,2)
  schepp: crooked (12,1)
die Schipp, Schibbe: spade (8,1)
  schicke: to send (5,1)
  schidde, gschitt: to pour (10,1)
  schier gaar: almost (3,2)
  schiesse, gschosse: to shoot
    (5,1)
 es Schild, -er: sign (11,1)
der Schildposchde: sign post (11,1)
die Schil(t)grodd, -e: turtle, tor-
    toise (9,2)
die Schindel, Schindle: shingle (6,1)
die Schissel, Schissle: bowl (6,1)
der Schitz: shooter, gunner (5,1)
  schkeede, gschkeedt: to skate
    (5,2)
die Schkeets (pl.): skates (5,2)
die Schkriendier, -e: screen door
    (6,1)
 es Schkwaer, -s: square (12,1)
  schleefrich: sleepy (10,2)
  schlaage, gschlaage: to hit
    (10,1)
  schlabbe: to slop (10,1)
  schlabbisch: sloppy, sloppily
    (11,1)
der Schlabbkiwwel: slop pail (10,1)
  schlachde, gschlacht: to slaugh-
    ter (8,1)
 es Schlachthaus, -heiser: slaughter
    house (8,1)
die Schlang, -e: snake (9,2)
  schlecht warre: to get sick (to
    stomach) (12,1)
 es Schleffelwedder: thawing weather
    (5,2)
  schleffle, gschleffelt: to thaw
    (5,2)
  schleiche, gschliche: to sneak
    (13,1)
  schleife, gschliffe: to grind
    (8,1)
der Schleifschtee: grindstone (8,1)
der Schlidde: sled (5,2)
  Schlidde faahre: to go sledding
    (5,2)
  schlippe: to slip (5,2)
  schlipperich: slippery (5,2)
der Schlissel, -le: key (6,1)
 es Schlisselloch, -lecher: key
    hole (6,1)
  schloofe, gschloofe: to sleep
    (6,1)
der Schloofkopp, -kepp: sleepyhead
    (10,1)
die Schloofschtubb, -e (-schtuwwe):
    bedroom (6,2)
die Schlooss, -e: hailstone (5,2)
 es Schloossewedder: hail storm (5,2)
der Schlupp, Schlubbe: tie (13,1)
  schmacke: to smell, taste (8,2)
  schmecke: to taste (8,2)
  schmeisse, gschmisse: to throw
    (5,2)
  schmelse, gschmolse: to melt
    (5,2)
  schmooke, gschmookt: to smoke
    (8,1)
 es Schmookhaus, -heiser: smokehouse
    (8,1)
der Schnee: snow (2,2)
der Schneeball, -e: snowball (5,2)
  schnee-e: to snow (5,2)

der Schneeflocke: snowflake (5,2)
die Schneeschaufel: snow shovel (5,2)
die Schneeschipp, -schibbe: snow
   shovel (5,2)
der Schneeschtarrem: snowstorm (5,2)
   schneide, gschnidde: to cut (10,1)
   schnell: fast, quickly (1,1)
die Schnitz (pl.): dried sections of
   apple, dried fruit (9,1)
die Schnook, -e: mosquito (10,2)
  es Schnuppduch, -dicher: handkerchief (13,1)
die Schnur: string (5,1)
der Schobb (Schubb), -e: shed (8,1)
  es Schockellied, -er: lullaby (10,2)
der Schockelschtuhl, -schtiehl: rocking chair (2,2)
   schockle, gschockelt: to rock
   (2,2)
der Scholle: clod (9,2)
die Schpaat, Schpaade: spade (9,1)
der Schpeck: bacon (8,1)
die Schpeckmaus, -meis: bat (10,2)
   schpeeder: later (3,2)
der Schpeicher: second floor (6,1)
   schpelle: to spell (1,2)
  es Schpellingbuch: spelling book
   (1,1)
schpende, gschpent: to spend (11,2)
der Schpicket, -s: spigot (6,1)
   schpiele: to play (1,2)
die Schpielbank, -benk: sink (6,1)
der Schpiellumbe: dishcloth (6,1)
  es Schpielsach, -e: toy (5,1)
die Schpielschissel, -schissle: dishwashing basin (6,1)
der Schpiggel, -e: mirror (6,2)
der Schpinaat: spinach (9,1)
die Schpinn, -e: spider (10,2)
   schpot: late (1,2)
die Schpring: spring (water) (8,1)
   schpringe, gschprunge: to "go together," date; to jump, run (2,2)
die Schpring(s)eeg, -e: spring harrow
   (9,2)
der Schpringer: a very young little
   fellow, baby boy (2,2)
  es Schpringhaus, -heiser: spring
   house (8,1)
der Schpringwagge, -wegge: spring
   wagon (8,2)
   schpritze: to squirt (9,1)
die Schprooch, -e: language (1,2)
der Schpruuss: spruce (9,2)
  es Schpuuck, -e: ghost, spook (5,1)
   schrecke, gschrocke: to scare (9,1)
der Schreibdisch, -e: desk (1,1)
   schreiwe, gschriwwe: to write (1,1)
die Schrottflint, -flinde: shotgun
   (5,1)
der Schtaab: dust (6,2)
die Schtadt, Schtedt: city (2,1)
die Schtadtleit (pl.): city people
   (2,2)
die Schtann (Schtaern): forehead (12,
   2)
der Schtann, -e: star (4,1)
der Schtarrem, -e: storm (5,2)
   schtarrick: strong, (fast) (5,1)
   schteche, gschtoche: to sting; to
   stick, prick (10,2)
  es Schteddel, -cher: village (1,1)
der Schtee: stone (9,2)
die Schteeck, -e: stairs (6,2)
die Schteefens, -e: stone fence (9,2)
die Schteemauer, -e: stone wall (8,2)

der (also die) Schteeschen, -s: station (12,1)
   schteh, gschtanne: to stand (1,1)
die Schteil, -s: style (13,1)
   schtelle: put, to place, set
   (3,2)
   die Uhr schtelle: set the
   clock (3,2)
der Schtengel, Schtengle: stalk (9,2)
  es Schtick, -er: piece (1,2)
die Schtieminschein, -e: steam engine
   (12,1)
   schtill: quiet (10,1)
der Schtiwwel, -e: boot (9,1)
der Schtock, Schteck: floor, story
   (of building) (6,1)
   schtoosse, gschtoosse: to churn
   butter; to bump, push (8,2)
   schtobbe: to darn (as a sock)
   (10,2)
die Schtorie, -s: story (12,1)
der Schtorkipper: storekeeper (11,2)
der Schtraahl,-e: ray, beam, stream
   (4,2)
   sich schtreehle: to comb oneself
   (10,1)
   schtricke, gschtrickt (schtrickle,
   gschtrickelt): to knit (10,2;
   13,1)
die Schtrooss, -e: street (6,1)
der Schtrump, Schtrimp: stocking
   (10,2)
die Schtubb, -e (Schtuwwe): room
   (1,1)
der Schtuhl, Schtiehl: chair (1,1)
die Schtunn, -e: hour (3,2)
   schtunnelang: for hours, hours
   long (4,1)
der Schtunnezoiyer: hour hand (3,2)
der Schubbkarrich: wheelbarrow (8,1)
die Schubblaad, -laade: drawer (6,1)
der Schuh: shoe (5,2)
der Schuhbendel: shoelace (5,1)
der Schuhmacher: shoemaker (11,2)
der Schuler, Schieler (also Schuler):
   pupil (1,2)
die Schulgleeder (pl.): school
   clothes (10,1)
der Schulhof, -heef: school yard
   (1,2)
  es Schunkefleesch: ham (8,1)
   schunn(t): already (1,1)
   schunnscht: otherwise (11,2)
   schuur: sure (5,2)
   fer schuur: for sure (5,2)
der Schuschder: shoemaker (11,2)
der Schwaerdaaddi: father-in-law
   (2,1)
die Schwaermammi: mother-in-law (2,1)
die Schwalm, -e: swallow (10,2)
der Schwamm, Schwemm: meadow (9,1)
   schwaz: black (4,1)
  es Schwazbord, -e: blackboard (1,1)
der Schweess: sweat (5,2)
   schwense: to play hooky, skip
   school (1,2)
   schwer: difficult, hard; heavy
   (2,2)
die Schweschder, -e: sister (1,2)
  es Schweschderskind, -kinner:
   nephew, niece (2,1)
   schwetze: to talk (1,1)
   schwiel: sultry (5,2)
   schwimme, gschwumme: to swim
   (5,1)
der Schwimmer: swimmer (5,2)

es Schwimmloch, -lecher: swimming hole (5,2)

schwitze: to sweat (5,2)

see-e, gseet: to sow (9,2)

die Seeg, -e: saw (8,1)

der Seegbock, -beck: saw horse (8,1)

die Seef: soap (6,2)

die Segund, -e: second (3,2)

sehne, gsehne: to see (2,2)

sei, gwest: to be (1,1)

es Seiche, -r: piglet (8,2)

der Seider: cider (9,1)

der Seischtall, -schtell: pig pen (8,2)

die Seit, Seide: side (2,2)

der Sellaat: lettuce, salad (9,1)

seller, selli, sell: that (1,2)

der Sellerich: celery (9,1)

selwer(t): self, oneself (8,1)

die Sens, -e: scythe (8,1)

der Sessel: easy chair (6,1)

sex(e): six (2,2)

die Sichel, Sichle: sickle (8,1)

es Siesses: sweets (11,2)

der (die) Singk, -s: sink (6,1)

die Sitti, -s: city (12,2)

sitze, gsotze (gsesse): to sit (1,1)

siwwe: seven (2,2)

der Siwweder: the number seven (13,1)

so: so, much (1,2)

es Sobber (Sabber), -s: supper (4,2)

der Socke: sock (13,1)

sogaar: even (2,2)

die Sohl, -e: sole (of shoe) (13,1)

solle, gsott: should (3,1)

es Soodbrenne(s): heartburn (12,1)

es Soofe, -s: sofa (6,1)

suche: to look for, seek (5,1)

es Suddelwedder: drizzly weather (5,2)

der Suh (Soh), Seh: son (2,1)

die Suhnsfraa (Sohnsfraa), -weiwer: daughter-in-law (2,1)

summerich: summery (5,2)

die Summerkich, -e: summer kitchen (6,1)

die Summerwa(er)scht: summer sausage (8,1)

die Sunn, -e: sun (4,2)

sunndaags: (on) Sundays (4,1)

die Sunndaagsgleeder (pl.): Sunday clothes (4,1)

die Sunndaagsschul, -e: Sunday School (4,1)

der Sunneschtich, -e: sun stroke (5,2)

der Sunneschtraahl, -e: sun ray (10,1)

sunnich: sunny (5,2)

die (es) Suut, -s: suit (13,1)

T

der Tarn, -e: tower, steeple (11,1)

der Tarreki, -s: turkey (5,1)

der Tietscher: teacher (m.) (1,1)

die Tietschern: teacher (f.) (1,1)

die Toilet, Toilets: toilet (6,2)

die Tomaet (Tomat), -s: tomato (9,1)

der Tschentelmann, -menner: gentleman (2,2)

U

uff: on, upon (1,1)

uff: up (1,2)

uffbasse, uffgebasst: to look out, be careful (6,1)

uffbinne, uffgebunne: to tie up (9,1)

uffbreche, uffgebroche: to break up (9,2)

uffgeh, uffgange: to go up (4,2)

uffheere, uffgheert: to stop (10,1)

uffhenke, uffghanke: to hang up (13,1)

uffhewe, uffghowe: to pick up; to lift up (8,2)

uffkoors: of course (1,2)

uffmache, uffgemacht: to open (6,1)

uffschliesse, uffgschlosse: to unlock (6,1)

uffschparre (-schpaerre), uffgschparrt: to open wide (12,1)

uffschteh, uffgschtanne: to get up (out of bed); to stand up (4,1)

uffsei, uffgwest: to be up (out of bed) (10,1)

uffwecke, uffgeweckt: to wake up (10,1)

die Uhr, -e: clock, watch (1,1)

es Uhregsicht: face of clock (3,2)

um: around (3,2)

um vier Uhr: at four o'clock

der Umberell: umbrella (5,2)

sich umdrehe: to turn (oneself) around (12,1)

es Umgraut (Ungraut): weeds (9,1)

un: and (1,1)

ungefehr: approximately, about (6,1)

es Unglick, -e: accident (11,2)

unne: without (6,1)

unne: beneath, below (1,1)

unnergeh, unnergange: to go under; set (sun) (4,2)

die Unnergleeder (pl.): underwear (13,1)

es Unnerhemm, -er: undershirt (13,1)

die Unnerhosse (pl.): underpants (13,1)

der Unnerrock, -reck: petticoat (13,1)

unnich: under (4,1)

unnersuche, unnersucht: to examine (12,1)

der Urenkel: great grandson (2,2)

der Urgroossdaadi: great grandfather (2,2)

V

es Vaddel: quarter (3,2)

der Vadder, Vedder: father (2,1)

der Vaddersdaag: Father's Day (5,1)

der Valentinsdaag: St. Valentine's Day (5,1)

vanne: (in) front (1,1)

vedderscht: front-most (6,1)

ver alders: long ago; in former times (9,1)

verbei: past, gone, over (4,2)

verbeigeh, verbeigange: to go by, go past (4,2)

verbeikumme, verbeikumme: to come by (6,2)

verdiene: earn (2,2)

verdollt: confounded (by) (10,2)

verdrickt: mashed, squashed (10,2)
verennere: change; to alter (as
   clothes) (13,1)
verheiert: married (2,1)
verhungere, verhungert: to starve
   (10,2)
verkaafe, verkaaft: to sell (8,2)
verleicht (villeicht): perhaps
   (2,1)
verliere, verlore: to lose (5,2)
sich verlosse, verlosse: to depend
   (10,1)
verschiddlich (verschiedlich):
   various (8,2)
sich verschloofe, verschloofe: to
   oversleep (10,1)
verschpreche, verschproche: to
   promise (12,2)
verschproche: "promised" (2,2)
sich verschteck(e)le, verschteck-
   elt: to hide oneself (10,1)
verschteh, verschtanne: to understand (1,1)
versuddle, versuddlet: to soil,
   ruin, make a mess of (13,1)
die Verwandte (pl.): relatives (2,2)
verzeehle, verzeehlt: to relate,
   tell (11,1)
viel(e): much, many (1,1)
vier: four (1,1)
villeicht (verleicht): perhaps
   (2,1)
der Voggel, Veggel: bird (6,2)
voll: full (4,2)
der Vorgang, -geng: foyer, front hall
   (6,1)
vorgeschder: day before yesterday
   (4,1)
der Vorhang, -heng: curtain (6,1)
der Vorschuss: overhang of a Swiss
   barn on the barnyard side (8,2)
vorsichtich: careful(ly) (11,1)
vun: from, of (1,2)

W

der Waage, Wegge: wagon (8,2)
waahr: true (5,2)
waarde: to wait (1,2)
waarm (warrem): warm (5,1)
wacker: awake (4,2)
wacker warre (waerre): to wake up
   (4,2)
waer?: who? (1,1)
der Waerdaag: weekday (4,1)
waerdaags: (on) weekdays (4,2)
die Waerdaagsgleeder (pl.): weekday
   clothes, work clothes (4,1)
waere, gewore: to wear (5,2)
der Waerschder (Warschder): sausage
   maker (8,1)
der Wahldaag: election day (5,1)
die Wand, Wend: wall (1,1)
es Wandbabier: wallpaper (6,1)
warre (waerre), geworre: to become
   (2,1)
es Wasser: water (5,2)
watsche: to watch (10,2)
es Watt, Waddej word (1,1)
es Wattshaus (Waertshaus): inn (11,1)
die Watzel (Wartzel), Watzle (Wartzle):
   root (9,1)
waxe, gewachst: to grow (5,2)
weck: away (8,1)
   weit weck: far away (8,1)

weckduh, weckgeduh: to put away
   (10,2)
wecknemme, weckgenumme: to take
   away (8,1)
weckschtelle, weckgschtellt: to
   put away (10,1)
es Wecksel: change (coin money)
   (12,2)
die Weckuhr, -e: alarm clock (10,1)
weckwesche, weckgewesche: to wash
   away (10,1)
weech: soft (9,2)
weege: because of, on account of
   (6,1)
weeich: because of, on account
   of (6,1)
der Weeze: wheat (9,2)
es Weezefeld, -felder: wheat field
   (9,2)
wedder: against (6,1)
es Wedder: weather (5,1)
der Wedderbericht: weather report
   (5,1)
der Wedderbrofeet, -eede: weather
   prophet (5,1)
weddere: to weather, to storm
   (5,2)
wedderleeche: to lighten (lightning) (5,2)
die Wedderrud: lightning rod (5,2)
der Weg, -e: road, way (11,1)
   sich uff der Weg mache: to
   start off (11,1)
es Weggeli, -n: little wagon (10,1)
weh: hurt (10,2)
es Weibsmensch, Weibsleit: woman
   (2,2)
der Weidebaam, -beem: willow tree
   (9,1)
weider: on, farther, further
   (3,1)
weidergeh, weidergange: to go on,
   proceed (3,1)
es Weifass, -fesser: wine barrel,
   cask (9,2)
die Weihnachde (pl.): Christmas (5,1)
die Weil: while (4,2)
   en Weil zrick: a while ago
   (4,2)
weise, gewisse: to show (6,2)
weiss: white (2,2)
weissle, geweisselt: to whitewash
   (8,1)
weit: far (5,2)
   weit fatt: far away (5,2)
der Welschhaahne: turkey gobbler
   (5,1)
es Welschhinkel: turkey hen (5,1)
es Welschkann: corn (9,2)
die Welschkanngribb: corn crib (8,2)
die Weschbohl: washbowl (6,2)
der Weschdaag, -e: washday (4,1)
wesche, gewesche: to wash (4,1)
   sich wesche, gewesche: to wash
   oneself (11,1)
es Weschheisel: wash house (8,1)
der Weschkarreb: washbasket (4,1)
die Weschlein: washline (6,2)
der Weschlumbe: washcloth (6,2)
die Weschmaschien, -e: washing
   machine (6,1)
die Weschp, Weschbe: wasp (10,2)
die Weschschpell, -e: clothespin
   (6,2)
der Weschzuwwer, -ziwwer: washtub
   (6,2)

wess(e)re: to water (9,2)
widder: again (2,1)
wie?: how? (1,1)
die Wieg, -e: cradle (6,2)
wiege, gewogge: to weigh (12,1)
wiescht: ugly (2,2)
wieviel, (wiffel): how many, how
  much (3,1)
der Wind: wind (5,1)
der Winder, -e: winter (1,2)
der Winderdaag: winter's day (5,2)
der Winderwind: winter wind (5,2)
windich: windy (5,2)
winsche: to wish (5,1)
die Wiss, -e: meadow (9,1)
wisse: to know (a fact) (1,1)
die Wissemaus, -meis: meadow, field
  mouse (9,2)
die Wittfraa, -weiwer: widow (2,2)
die Woch, -e: week (4,1)
wochelang: for weeks (5,2)
die Wohning, -e: dwelling, residence
  (2,1)
die Wohnschtubb, -e (-schtuwwe): living room (5,1)
die Wolk, -e: cloud (5,2)
die Woll: wool (13,1)
woll: woolen (13,1)
woll (wull): probably, no doubt
  (12,2)
wolle (welle), gewollt: to want
  to (5,1)
der Wuffi: dog's name (6,1)
wu?: where? (1,1)
wuhie?: where to? (4,1)
die Wuhnschtubb,-e (-schtuwwe): living room (6,1)
wunnerbaar: wonderful(ly) (2,1)
der Wunsch, Winsch: wish (5,1)

                    Y

 es Yaahr, -e: year (3,1)
yaage, geyaagt: to hunt (chase)
  (5,1)
die Yaahreszeit, -zeide: season (5,2)
yeder: every, everyone (5,1)
yederebber: everyone (1,2)
yetz(t): now (12,1)
yo: yes (to a negative question);
  particle used for emphasis (4,1)
die Yunge (pl.): young (of animals)
  (8,2)
yuscht: just (1,1)

                    Z

der Zaah, Zeeh: tooth (12,1)
der Zaahdokder, (Zeehdokder): dentist
  (12,1)
 es Zaahfleesch: gum(s) (12,2)
zamme: together (1,2)
zeehle: count (3,1)
zehe: ten (1,2)
 es Zehrgeld: spending money (12,2)
 es Zehuhrschtick: food eaten during
  a pause in the morning's work
  (10,1)
zeidich: ripe (9,2)
zeidiche: to ripen, mature (9,2)
die Zeiding, -e: newspaper (4,2)
z(e)rick: back (4,2)
die Zeit, Zeide: time (1,2)
der Zent, (Sent): cent (2,2)

zidder: since (6,1)
der Zidderli: souse (8,1)
ziehge, gezogge: to pull, move
  (2,2)
der Ziffer, -e: number, figure (3,2)
 es Zifferblatt (blaat): dial (face)
  of clock or watch (3,2)
zimmlich: quite, rather (2,1)
der Zoiyer: hand of watch or clock
  (3,2)
zu: to, too (1,2)
zu: closed (1,2)
der Zuckerschtengel, -le: candy
  stick (11,2)
zudrehe: to turn off (6,1)
zufridde: satisfied (10,1)
zugucke, zugeguckt: to observe,
  look at (6,1)
zumache, zugemacht: to close
  (6,1)
zuneehe, zugeneeht: to sew shut
  (13,1)
die Zung, -e: tongue (12,1)
zwee: two (2,1)
zweeschteckich: two-storied, two
  floors high (6,1)
zwelf(e): twelve (2,2)
zwett: second (2,1)
der Zwilling,-e: twin (2,2)
zwische: between (6,1)
die Zwiwwel, Zwiwwle: onion (9,1)

APPENDIX

SAMPLES OF PRACTICE PATTERNS

Chapter One, Part One

Practice patterns are to the language student what
scales and etudes are to the music student; just as the
music student practices his scales so that he can play
his melodies, songs, concertos, and the like, with
greater confidence and thus more smoothly, just so you,
the language student, should practice the following
patterns. Like scales, the patterns aren't much fun,
but they will give you a facility and confidence in
conversational PG that you could otherwise acquire only
through hours of conversation.

You should first of all familiarize yourself with
the reading selections by reading them several times.
After--or even during--the second reading, begin asking simple questions about the material--the subject,
verb, objects--in each sentence. Your teacher will
help you develop this technique, of course, but here
are some interrogatives, or question words, to get you
started:

Was?   What?
Was macht ... ? What is ... doing?
Waer?   Who?
Wu?   Where?
Wuhie?   Where to?
Wann?   When?

Now for a few examples to show how this technique is
applied to the first reading selection.

We read: Die Schulbell ringt.
    We ask: Was ringt?
We answer: Die Schulbell ringt.
    We ask: Was macht die Schulbell?
We answer: Die Schulbell ringt.

We read: Die Kinner gehne ins Schulhaus.
    Ask: Waer geht ins Schulhaus?
Answer: Die Kinner gehne ins Schulhaus.
    Ask: Wuhie gehne die Kinner?

293

Answer: Ins Schulhaus gehne die Kinner.
Ask: Was mache die Kinner?
Answer: Sie gehne ins Schulhaus.

Silly? About as silly as practicing scales. The
trick, of course, is to make your questions and
answers sound as conversational as possible. Again,
follow the lead of your instructor. If you have a
friend or a family member who knows PG or is learning
it too, take turns asking and answering the questions.

Pronouns

Now a variation to practice pronouns follows very
naturally:

Ask: Ringt die Schulbell?
    Answer: Ya, sie ringt.
Ask: Hot es Schulhaus ee Schtubb?
    Answer: Ya, es hot.
Ask: Sitzt der Schulmeeschder am Schreibdisch?
    Answer: Ya, er sitzt am Desk (Dest).

Use this same practice pattern as you look at the
vocabulary list; use the various nouns that could possibly be the subjects of the various verbs in the list.

Verbs

Once again, you want the following practice patterns
to sound just as natural as possible. First of all,
look at and study the verb kumme. After you have
familiarized yourself with the conjugation, you can
start in:

Kummscht du?
Ya, ich kumm, un aer kummt aa. Mer kumme all.
Kummt ihr (kummt [d]ihr) aa?
Gewiss. Mir kumme un sie kumme aa.

Make believe you are actually talking to someone.
Emphasize subjects or verbs as you would if really asking and answering questions: Are you coming? Sure,

I'm coming, and he's coming too. We're all coming!

Then you can begin adding other bits and pieces of material:

Gehscht du in die Schul?
Gewiss. Ich geh un sie geht aa. Mer gehne
all in die Schul.
Geht dihr aa in die Schul?
Ya, mer gehne in die Schul, un sie gehne aa.

Ich kann Deitsch schwetze. Kannscht du?
Nadierlich kannich (kann ich); mer kenne all
Deitsch schwetze.
Kennt dihr Deitsch schwetze?
Gewiss. Mir kenne un sie kenne aa.

Practice the verbs in the grammar section and in the vocabulary lists in this manner, and you will soon gain a facility and confidence that you would never have otherwise, even if you were to read the text several times or just memorize the vocabulary list.

Chapter Three, Part One

The Adjective and the Pronoun eens

In the following pattern note the relation between ken(s) and eens.

Hoscht (du) en Onkel? (m.)
Nee, ich hab kenner. Hoscht du eener?
Ya, ich hab yuscht ee Onkel.

Hoscht du en Ant? (f.)
Nee, ich hab kenni. Hoscht du eeni?
Ya, ich hab ee Ant.

Hoscht du en Kind? (n.)
Nee, ich hab kens. Hoscht du eens?
Ja, ich hab yuscht ee Kind.

Replace <u>Onkel</u>, <u>Ant</u>, and <u>Kind</u> with appropriate nouns
from the vocabulary.

This next pattern will help you to practice not only
the dative forms of <u>eens</u>, but also the plurals of nouns,
and the verbs <u>hawwe</u> and <u>gewwe</u> (+ dative).

Ich hab zwee Brieder. Eem Bruder gewwich en
Buch, un eem gewwich en Bensil.

Sie hot zwee Schweschdere. Eenre Schweschder
gebt sie zehe Zent, un eenre gebt sie zwansich Zent.

Mer hen zwee Kinner. Eem Kind gewwe mer en
Fedder, un eem gewwe mer en Schtick Babier.

Now pick appropriate words from the vocabulary list
of this and the previous chapters.

SAMPLES OF MINIMUM ESSENTIALS

Chapter One, Part One

1. The Definite Article, Nominative and Accusative
   Cases
2. The Indefinite Article, N. and A. Cases
3. Personal Pronouns, N. and A. Cases
4. Verbs
   kumme, to come
   saagge, to say
   schreiwe, to write
   geh, to go
   sei, to be
   wisse, to know
   kenne, to be able to, can
   kenne, to be acquainted with

And the vocabulary to translate the following sentences:

1. Good morning (good evening), how are you?   (how
   goes it?)
2. Good, thank you, and how are you?
3. The teacher (masc.) asks, "Who are you?"
4. The pupil says, "I am _____."
5. He asks, "What is your name?"
6. My name is _____.
7. The teacher (fem.) speaks English and German.
8. Can you speak German?
9. Yes, I can. I can also read and write.
10. And where do you live?
11. Some children live in the country.
12. Others live in (the) town.
13. Are you smart?
14. Yes, I know everything.
15. Some pupils are dumb (stupid) and know nothing.

NOTE: Translations of the English sentences are always
       placed on the back of the hand-outs.

1. Gud(e) Marriye (Guden Owet), wie bischt?   (wie
   geht's?)
2. Gut, danke, un wie bischt du?
3. Der Meeschder (Tietscher) froogt, "Waer bischt du?"

4. Der Schuler saagt, "Ich bin der _____."
5. Er froogt, "Was iss dei Naame?   (Wie heescht du?)"
6. Mei Naame iss _____.
7. Die Tietschern (Meeschdern) schwetzt englisch un
   deitsch.
8. Kannscht du deitsch schwetze?
9. Ya, ich kann. Ich kann aa lese un schreiwe.
10. Un wu wuhnscht du?
11. Deel Kinner wuhne uff em Land.
12. Annere wuhne im Schteddel.
13. Bischt du gscheit?
14. Ya, ich weess alles.
15. Deel Schuler (Schieler) sin dumm un wisse nix.

Chapter Two, Part Two

1. Demonstratives
2. ken, kee
3. Verbs
     leige, to lie (recline)
     draue, to trust
     helfe, to help

And the vocabulary to translate the following sentences:

1. My parents have a son and a daughter.
2. She (accented demonstrative) is my sister, and he
   (demonstrative) is my brother.
3. Don't you have no uncle? No, I have none.
4. He is lying on the bed.   (Replace he with other
   pronouns.)
5. I like the baby. (Replace I.)
6. My father's mother is still living.
7. I like her cooking and baking.
8. Do you trust him? (Replace you and him.)
9. No, him (demons.) I don't trust.   (Replace him and
   I.)
10. Don't you have no aunt? No, I have none.
11. I'm helping my father.   (Replace I and my; use
    mother.)
12. Don't you trust that woman?
13. No, her (demons.) I don't trust!
14. Is he helping those people?

15. No. he's not helping them (demons.).

 1. Mei Eldre hen en Suh (Soh) un en Dochder.
 2. Die iss mei Schweschder un daer iss mei Bruder.
 3. Hoscht du ken Unkel? Nee, ich hab kenner.
 4. Er leit uffem Bett.
 5. Ich gleich es Bobbel.
 6. Meim Vadder sei Mudder iss noch am Lewe.
 7. Ich gleich ihre Koches un Backes.
 8. Drauscht du ihm?
 9. Nee, dem drau ich net.
10. Hoscht du net kee Aent? Nee, ich hab kenni.
11. Ich helf meim Vadder (meinre Mudder).
12. Drauscht du net daerre Fraa?
13. Nee, daerre drau ich net.
14. Helft aer denne Leit?
15. Nee, denne helft er net.